OXFORD REVISION GUIDES

GCSE

Design and Technology

TEXTILES TECHNOLOGY
through diagrams

Jane Down

OXFORD
UNIVERSITY PRESS

OXFORD
UNIVERSITY PRESS

Great Clarendon Street, Oxford OX2 6DP

Oxford University Press is a department of the
University of Oxford. It furthers the University's objective
of excellence in research, scholarship, and education by
publishing worldwide in

Oxford New York
Auckland Bangkok Buenos Aires Cape Town Chennai
Dar es Salaam Delhi Hong Kong Istanbul Karachi Kolkata
Kuala Lumpur Madrid Melbourne Mexico City Mumbai Nairobi
São Paulo Shanghai Singapore Taipei Tokyo Toronto

with an associated company in Berlin

Oxford is a registered trade mark of Oxford University Press
in the UK and in certain other countries

British Library Cataloguing in Publication Data

data available

ISBN 0 19 832834 6

10 9 8 7 6 5 4 3 2

Typeset and designed by Hardlines, Charlbury, Oxon

Printed and bound in Great Britain

CONTENTS

Specifications for GCSE Design and Technology: Textiles Technology

GCSE Textiles Technology comes under the heading of Design and Technology. It may be studied as a full or short course. Both courses can be assessed at a foundation or higher tier (level). The grades available for each tier are shown below:

	Tier	Grades available
GCSE Textiles Technology full and short courses	Higher	A*–D
	Foundation	C–G

EXAMINING GROUPS AND SPECIFICATIONS

There are several examining groups or boards that set specifications and examinations. Make sure you know which group and specification you are following. They may differ in coursework requirements, length of examination papers, etc. The content of all Textiles Technology specifications is broadly the same because they must follow the National Curriculum for Key Stage 4. Therefore this revision guide will help all GCSE Textiles Technology students, regardless of the specification being studied.

Examination group/ specification	Course type	Tier	Length of paper	Coursework	Assessment		
AQA (formerly SEG, Southern Examining Group)	Full 3547	Higher	2 hours	Design and Make Project 40 hours	Coursework	Exam	Total
		Foundation	2 hours		60%	40%	100%
		Preparation sheet issued in March, giving the design context for design questions.					
	Short 3557	Higher	$1\frac{1}{2}$ hours	Design and Make Project 20 hours	60%	40%	100%
		Foundation	$1\frac{1}{2}$ hours				
		Preparation sheet issued in March, giving the design context for design questions.					
EDEXCEL London Examinations	Full 1971	Higher	Paper 2H $1\frac{1}{2}$ hours	Design and Make Project 40 hours	Coursework	Exam	Total
		Foundation	Paper 2F $1\frac{1}{2}$ hours		60%	40%	100%
	Short 3971	Higher	Paper 2H 1 hour	Design and Make Project 20 hours			
		Foundation	Paper 2F 1 hour		60%	40%	100%
OCR (formally MEG, Midland Examining Group)	Full 1958	Higher	**2** 1 hour 15 mins **4** 1 hour 15 mins	Design and Make Project 40 hours	Coursework	Exam	Total
		Foundation	**1** 1 hour **3** 1 hour		F **1** 20% H **2** 20% F **3** 20% H **4** 20% Total 60%	40%	100%
	Short 1058	Higher	$1\frac{1}{2}$ hours	Design and Make Project 20 hours	60%	40%	100%
		Foundation	1 hour				

Designing and making skills

GCSE Textiles Technology provides opportunities to use Textiles for investigation, designing, making and evaluation. In order to develop designing and making skills you must ensure you have a sound knowledge and understanding of the properties of fabrics. It is important that you know how the properties of fabrics, the processing techniques and the equipment and tools you choose affect the quality and finish of the final product and your Textiles Technology project!

The three Assessment Objectives are:

AO1 Capability through acquiring and applying knowledge, skills and understanding of materials, components, processes, techniques and industrial practice.

AO2 Capability through acquiring and applying knowledge, skills and understanding when designing and making quality products.

AO3 Capability through acquiring and applying knowledge, skills and understanding when evaluating processes and products and examing the wider effects of design and technology on society.

COURSEWORK INFORMATION

Different specifications have different coursework requirements. Make sure you understand **exactly** what you are expected to do before you begin.

Group/ specif- ication	Coursework requirements
AQA	Full: 40 hours Short: 20 hours Coursework will be internally assessed and externally moderated. Short and Full: The project should address all three assessment objectives in an integrated way. Candidates are required to submit a concise design folder and/or the appropriate ICT evidence and a 3-dimensional outcome. Throughout the project candidates should address the industrial and commercial practices, and the moral, social, cultural and environmental issues, arising from their work.
EDEXCEL	Full: 40 hours Short: 20 hours Coursework will be internally assessed and externally moderated. Short: Coursework project covers all aspects of designing and making, AO2. Centres are advised to use the 10 page A3 proforma available from Edexcel. The project portfolio or a hard copy of equivalent ICT evidence must show, in summary form, student's designing and making skills. Full: Portfolio and product – no more than 40 hours. A single Design and Make task and portfolio. The task will be chosen by the student and approved by the teacher, who must make sure that the task will cover the three assessment criteria. The teacher may make modifications to the student's design proposal for safety or other reasons, provided the help given is recorded in the student's folio of work. Students who do not complete all aspects of the coursework, but show full coverage of all assessment criteria, will not be disadvantaged. The portfolio should consist of approximately 15 pages in an A3 format or a hard copy of the equivalent ICT evidence.
OCR	Full: 40 hours Short: 20 hours Coursework will be internally assessed. Short and Full: Candidates are expected to design and make a quality Textiles product. The project can be linked to candidates' own interests, industrial practice or the community. Projects may involve an enterprise activity and candidates must use appropriate ICT to help with their work. This can include computer-aided design and manufacture (CADCAM) software, control programs, data analysis and ICT-based sources for research. Candidates must consider how technology affects society and their own lives.

Preparing for revision

The key to getting through exams successfully is preparing yourself through organization and revision.

ORGANIZATION – WHAT SHOULD I REVISE?
Be sure about the content of the syllabus, either by obtaining a copy from your teacher or by sending for your own (see below).

WHAT CAN I EXPECT THE EXAM TO BE LIKE?
- Make sure you know the number and length of papers you have to sit (see page iv). Whichever examination you are sitting, the written exam will account for 40% of the marks.

- Are you going to be given a preparation sheet, a theme, or pre-release material prior to the examination? Check with your teacher.

GETTING UNDERWAY WITH YOUR REVISION!
- Start as early as possible – at least 3 months before the examination. Complete your coursework by then if possible so that you can concentrate on your revision.

- Make a revision timetable. USE IT!

BE PREPARED!

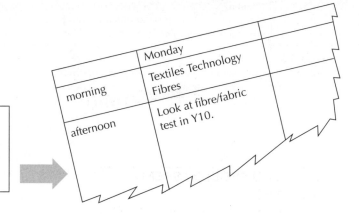

	Monday	
morning	Textiles Technology Fibres	
afternoon	Look at fibre/fabric test in Y10.	

Syllabus contacts

- **AQA**
  Stag Hill House, Guildford, Surrey GU2 7XJ
  Tel: 01483 506506 Fax: 01483 300152
  Devas Street, Manchester, M15 6EX
  Tel: 0161 9531180 Fax: 0161 2737572
  www.aqa.org.uk

- **EDEXCEL**
  Edexcel Publications
  Adamsway, Mansfield, Notts. NG18 4FN
  Tel: 01623 467467 Fax: 01623 450481
  www.edexcel.org.uk

- **OCR Information Bureau, General Qualifications**
  (Certificate of Achievement, GCSEs, AS and A Level, and the new GNVQ)
  Tel: 01223 553998 Fax: 01223 552627
  www.ocr.org.uk

MAKING A START?

- Start by organising yourself. To revise thoroughly and effectively you will need the following:

HOW LONG SHOULD I SPEND REVISING?

- There are no hard and fast rules as we are all individual in our needs. However 1 hour sessions are usually most effective, using 40 minutes to read and retain information and a further 20 minutes going back over your notes.
- Break the work into topics and tick off each topic as you complete it.

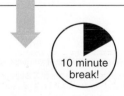

10 minute break!

DO:

- Choose a quiet place with plenty of light. TV and music distract you.
- Timetable in regular short breaks and stick to them.
- Remember! The exams won't last for ever!

THIS BOOK IS HERE TO HELP YOU!

- The information in this book covers the main content of Textiles Technology courses. It is designed to remind you of topics you have already covered in your lessons. The diagram format should help you remember key facts and information.

THIS BOOK IS NOT ENOUGH!

- Reading it is not enough!
  To revise and remember a topic you must be active!

Make notes. Use headings. Write down key words.

Use a highlighter pen on key words or facts

spider diagrams a useful way to remember facts about a topic

Give yourself a test!

$\frac{10}{10}$

Help! If you do not understand something ask your teacher!

Keep going back over your notes...

REPETITION AIDS MEMORY

Success in the examination

One of the main reasons candidates lose marks in an examination is because they mis-read instructions and questions. To prevent yourself from falling into this trap:

 When you turn over your paper, take time to read all the instructions carefully.

 Make a very careful note of how many questions you must answer in each section of the paper.

 Check back and front of each page to make sure you have seen all the questions and all the sections.

 Stay calm – do not panic! Read the questions carefully so that your brain registers the topics fully.

Read each question through at least twice before answering it. Make sure you know **exactly** what you are being asked.

Start by answering the questions where you are confident you know the answer; this will help you to relax.

Each question is awarded marks and these marks are highlighted clearly alongside each question. Spend a longer time on the questions with the highest marks.

If you find a question difficult, do not waste time on it at the start but always come back to it when you have time at the end.

TIME TO SPARE – DON'T WASTE IT!

If you find that you complete the paper before time, check over all your answers again carefully. Marks are given for punctuation and grammar, so check these as well. The second time around may bring a correct term to mind that you just couldn't think of at first. Checking always gains you a few valuable marks which can mean the difference between grades.

EXAMINATION QUESTIONS

Different types of question are used in examinations. Below are some examples:

Short-answer questions

Typical example:
Name two natural animal fibres.   (2 marks)

Answer:
Wool and silk.

Sometimes short-answer questions require you to complete a chart.

Open-ended questions

Typical example:
Explain micro-encapsulation in modern fibres.   (3 marks)

Answer:
Modern microfibres are produced with hollow centres which can be filled with beneficial chemicals and vitamins as tiny crystals. The crystals break down slowly and the fibres withstand prolonged wear and washing so the effects can be long lasting. Originally they were developed for space travel but are now used, for example, in sports clothing.

Other open-ended questions might start with 'describe', 'outline', or suggest'.

Structured questions
Typical example:
a)   Name the source of wool. [*Answer:* Sheep.]   (1 mark)
b)   Explain three main differences between fine wool and coarse wool fibres. [*Answer:* Fine wool fibres are shorter in length (4–7 cm) while coarse wool fibres can be up to 35 cm in length; the crimp on fine wool fibres is 12 loops per cm, and only 2 loops per cm on coarse wool; fine wool fibres have a low lustre while coarse wool fibres have a high lustre.]
                                                                                                                                       (3 marks)

c)   Wool fibre is very elastic. Describe why a wool fibre is elastic. [*Answer:* Wool fibre is made of long protein bundles which in warm conditions move apart as the fibre absorbs moisture, breaking down the bonds between the bundles. The bonds reform as the fibre dries and cools.]
                                                                                                                                       (3 marks)

Structured questions usually build up in difficulty so that marks can be gained by those candidates only able to answer the first parts of the question. Always answer any parts you can because you gain marks that way, even if you cannot attempt more difficult parts of the question.

Fibres

CLASSIFICATION OF FIBRES

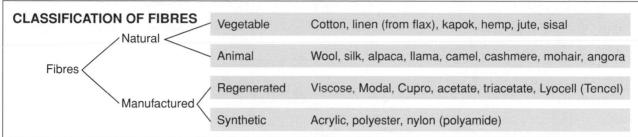

Fibres
- Natural
  - Vegetable — Cotton, linen (from flax), kapok, hemp, jute, sisal
  - Animal — Wool, silk, alpaca, llama, camel, cashmere, mohair, angora
- Manufactured
  - Regenerated — Viscose, Modal, Cupro, acetate, triacetate, Lyocell (Tencel)
  - Synthetic — Acrylic, polyester, nylon (polyamide)

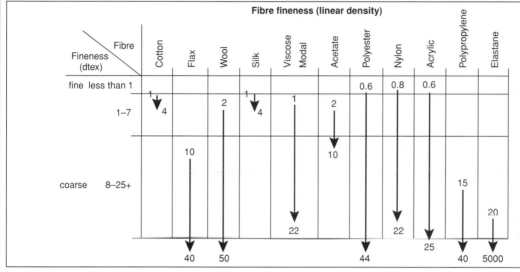

Fibre fineness (linear density)

Linear density, or **titre**, is the fibre weight per unit length. The units are **tex** or **dtex** (**decitex**). The smaller the number, the finer the fibre. For example, a fineness of 2 dtex indicates that 10 km of fibre has a mass of 2 grams.

PROPERTIES OF FIBRES

- affects the weight of the fabric made from the fibre: low-density fibres produce light, voluminous (bulky) fabrics
- shows how well a fibre absorbs water from the air, which also depends on the humidity (water content) of the air; good water-absorbing fibres help to prevent a build-up of static electricity in clothes
- the force needed to break the fibre – strong fibres have a high tenacity value; fabrics made from these fibres are very durable
- the shorter the fibre, the more fibre ends and hairiness to help trap air
- resistance to being decomposed by the enzymes of micro-organisms
- the distance the fibre will stretch before breaking and how well it returns to its original length after stretching
- fibre weight per unit length

Fibre	Fineness	Length	Density	Moisture absorption	Biological resistance	Tenacity		Breaking extension		Elasticity
						Dry	Wet	Dry	Wet	
Cotton	medium	staple	high	good	poor	high	high	low	higher	poor
Flax	coarse	long staple	high	v. good	poor	high	high	lowest	higher	poor
Wool	fine/coarse	staple	medium	v. good	poor	medium	lower	high	higher	good
Silk	medium	filament	medium	good	poor	high	lower	medium	higher	v. good
Viscose	fine/coarse	filament	high	v. good	poor	medium	much lower	medium	higher	poor
Acetate	medium	filament	medium	good	good	medium	lower	high	higher	good
Polyester	fine/coarse	staple	medium good	v. poor	v. good	high	lower	medium	medium	v. good
Nylon (polyamide)	fine/coarse	staple	low	poor	v. good	high	lower	high	high	v. good
Acrylic	fine/coarse	staple	low	v. poor	v. good	high	lower	medium	medium	v. good
Polypropylene	fine/coarse	staple	lowest	poorest	v. good	high	high	high	high	good
Elastane	coarse	filament	low	v. poor	good	low	lower	v. high	v. high	highest

Vegetable fibres: cotton and linen

COTTON

Cotton boll

Cotton plants need **two** conditions for growth:
- tropical climate
- wet soil

Each seed boll holds up to 30 seeds. Each seed is surrounded by up to 10 000 hairs or fibres.

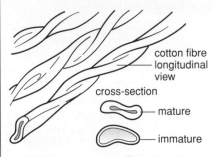

cotton fibre longitudinal view

cross-section

mature

immature

Cotton fibres

The round fibre dries to a kidney-shaped cross-section and twists along its length. The surface is wax covered.

THE COTTON-MAKING PROCESS

Harvesting
Plants are harvested by a fast cotton-picking machine.

↓

Drying
The wet bolls are dried by blowing warm air over them.

↓

Ginning
This process removes all the fibres from the seeds.

↓

Combing and carding
The fibres are combed and carded to make them lie in the same direction.

↓

Spinning
Cotton fibres are spun into staple yarns.

Cotton fibres are:
- poor insulators
- stronger when wet than dry
- able to hold moisture well
- hardwearing

Lint fibres are 15–50 mm in length.
Linters are shorter fibres which are often used to blend with other fibres for interfacings, cleaning cloths, etc.

INTERNATIONAL COTTON EMBLEM

Cotton emblem

LINEN

A flax plant

Flax plants grow well in a climate that is:
- cold
- damp

The linen fibres are extracted from the long stems of the plant.

The smooth linen fibres are:
- poor insulators
- highly absorbent
- hardwearing
- The lustre helps to prevent soiling

 linen fibre

 cross-section of linen fibre bundle

Linen fibres

THE LINEN-MAKING PROCESS

Pulling
A puller machine lifts the flax stems and lays them out on the fields.

↓

Roughing out
Unwanted plant materials are removed from the stems as they are collected.

↓

Retting
Micro-organisms are encouraged to grow on the stems by soaking them for a period in warm water. Their enzymes break down the stem to release the fibres.

↓

Breaking and scutching
This removes the outer stem parts, leaving flax fibres 45–90 cm long.

↓

Hackling
The best fibres are combed into spinnable fibre bundles. Remaining stem parts and short fibres are removed. Long fibres are called **line tow**; short fibres are called **hackle tow**.

↓

Spinning
Line tow fibres are spun into yarns for linen production.

LINEN SEAL

Linen Seal

Animal fibres: wool and silk

Natural animal fibres are formed from proteins. The most common are wool fibres, which come mainly from sheep, and silk fibres produced by silk worms. Silk fibre is the only naturally produced continuous filament fibre.

WOOL

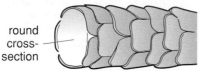

round cross-section

Wool fibre cross-section

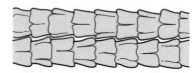

Interlocking effect of wool fibre scales when rubbed together

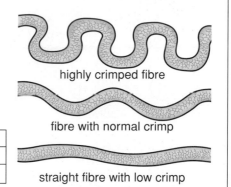

highly crimped fibre

fibre with normal crimp

straight fibre with low crimp

Fibre differences	Length	Crimp	Lustre (shine)
Fine wool fibre	4–7.5 cm	12 loops per cm	Low
Coarse wool fibre	35 cm	2 loops per cm	HIgh

Wool fibre production flow chart

Shearing
Sheep are shorn every year using electric shears. Care is taken to keep the fleece intact.
▼
Classing
Fleece is graded on fineness, crimp, length, impurities and colour into four qualities (1–4; 1 being best).
▼
Scouring
Gentle scouring removes grease (lanolin), dirt and burs. Lanolin is a natural oil secreted from the skin of a sheep to keep its fleece waterproof.
▼
Carbonizing
Some fleece contains a lot of plant material. Sulphuric acid is used to remove it from the fibres.
▼
Carding
Wire brush rollers disentangle the fibres to produce an even web of wool fibres (carded slivers).
▼
Gilling*
The slivers are pulled through coarse combs to align the fibres.
▼
Combing
Finer combs remove the short fibres (noils).
▼
Dyeing
Colour can be added to the loose wool here, or later, to the yarn or the woven fabric.
▼
Drawing*
A further gilling reduces the silver to a fine roving (a ribbon of fibres), prior to spinning.
▼
Spinning
The fibres are drawn out, twisted and wound onto reels as yarn. The yarn is used to make knitted or woven fabric (see page 9).

* stage required to produce fine worsted cloth

SILK

Silk production flow chart

Silk thread
Comes from a spinneret below the mouth of the caterpillar, forming a cocoon.
▼
Harvesting
The cocoons are harvested carefully so as not to damage them.
▼
Heat
Steam heat is used to kill the developing pupae inside the cocoons. Hot water is also used to soften the gum holding the cocoon together.
▼
Spinning
7 to 10 silk filament ends are wound together to form raw silk yarn.
▼
Silk reeling
The finished reel of yarn is 1000 metres long. Several yarns are twisted together for greater strength.

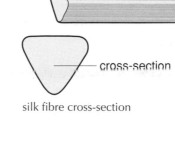

cross-section

silk fibre cross-section

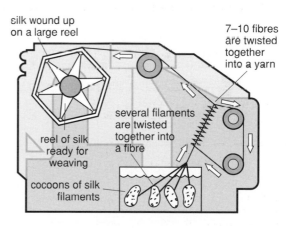

silk wound up on a large reel

7–10 fibres are twisted together into a yarn

several filaments are twisted together into a fibre

reel of silk ready for weaving

cocoons of silk filaments

PURE NEW WOOL

Wool Mark

Silk Seal

Chemical fibres

Definitions
Regenerated fibres are made chemically by changing natural materials such as wood pulp into chemical fibres.
Synthetic fibres are made from chemicals that come from oil and coal.

POLYMERS

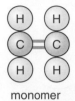

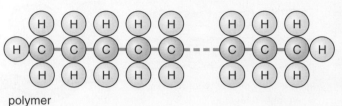

monomer   polymer

All fibres are made from chemical units called **polymers**, which are formed from single units called monomers linked together to form long chains.

Polymerization is the process of forming polymers.

VISCOSE – A REGENERATED FIBRE

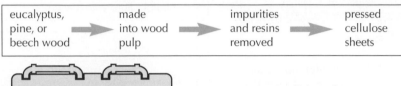

eucalyptus, pine, or beech wood → made into wood pulp → impurities and resins removed → pressed cellulose sheets

1 solvation

cellulose sollution

2 filtering

pump

3 extrusion

spinneret

4 spinning bath

5 drawing

6 washing

7 lubrication

8 drying

9 winding

filament yarn

1 Pressed cellulose sheets are dissolved in an organic solvent to form cellulose solution or viscose.

2 The solution is filtered.

3 The solution is forced through small holes in a spinneret (extruded).

4 The extruded fibres pass out into a bath of chemicals which harden the fibres as they emerge.

5 The fibres are drawn (stretched) out and gathered into a filament yarn.

6 The yarn is washed to remove chemicals.

7 The yarn is lubricated to make it supple.

8 The yarn is dried.

9 The yarn is wound onto a spool or cut into staple fibres.

MAIN TYPES OF SYNTHETIC FIBRE (CLASSIFIED BY CHEMICAL PROCESS)

- **Addition polymerization**
  Polymer unit: long polymer made from the same basic unit.
  Fibres produced by the process:
  acrylic
  polypropylene
  PVC (polyvinylchloride)
  PTFE (polytetrafluoroethylene)
  PVA (polyvinyl acetate)

- **Condensation polymerization**
  Polymer unit: long polymer made from the same basic unit.
  Fibres produced by the process:
  polyester
  nylon (polyamide)

- **Block co-polymerization**
  Polymer unit: long polymer made from alternating blocks of different units.
  Fibre produced by this process:
  elastane (Lycra)

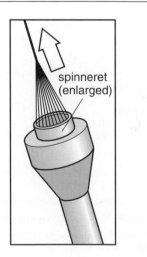

spinneret (enlarged)

Synthetic fibres

NYLON

Nylon is made from polyamide.

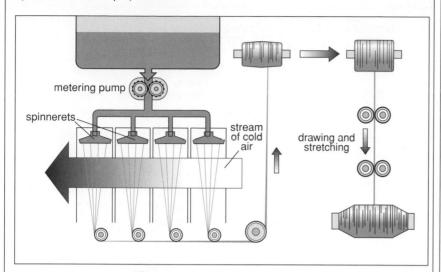

metering pump

spinnerets

stream of cold air

drawing and stretching

General properties of nylon:

- Absorbs little water except when texturing provides capillary action

- Very strong and resistant to wear

- Very crease resistant

- Very resistant to alkalis, and to many solvents and moulds, but can be attacked by acids

- Affected by static electricity which attracts dirt and affects drape

Monofilament
– single filament used for sewing thread.

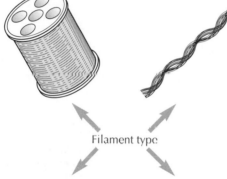

Filament type

Filament
Many filaments twisted together, usually textured and reinforced used for hosiery, lingerie, sportswear, household fabrics and carpets.

Staple fibres
made by chopping filaments – blended with other fibre types.

Aramid used in fibre-reinforced plastics such as Kevlar for protective clothing.

POLYESTER

General properties of polyester:

- Strong, hard wearing and crease resistant

- Unaffected by most acids, alkalis and solvents but can be degraded by very concentrated solutions of acids and alkalis

- Easy to wash and care for

- Resistant to staining

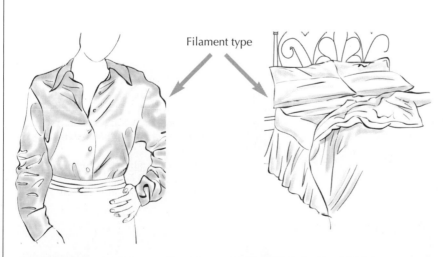

Filament type

Filament fibres usually textured: used for dresses, blouses, scarves, linings and net curtains

Staple fibres sometimes blended with other fibres: used for garments, bedlinen and wadding

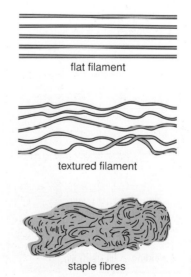

flat filament

textured filament

staple fibres

Microfibres

Microfibres are now being used by the fashion industry for a number of different purposes.

Example: Tactel – made from Polyamide 6.6 (6.6 refers to the chemical make-up of the fibre)
Uses:

  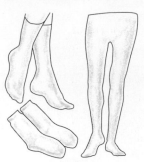 

Sportswear – 30% lighter than cotton, absorbs moisture readily, protects from wind, weather and low temperatures, not bulky or heavy, can be double knitted with inner layer of cotton.

Underwear – Absorbs moisture readily from the body. keeping it dry. It is soft, lightweight, lustrous, and has good draping qualities.

Hosiery – Extremely soft and luxurious, special fibre cross-section gives added lustre, lightweight.

Rainwear – Water repellent, breathable, keeps body at a constant temperature.

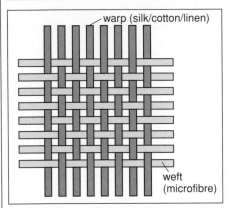

Cotton warp with a microfibre weft

Natural fibres, cotton, linen or silk are often used for the warp to reduce costs.

Micro-encapsulation

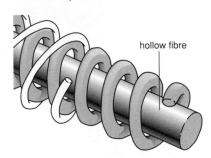

Micro-encapsulation within a fibre

Beneficial chemicals and vitamins are released slowly from the hollow fibre core to be absorbed through the skin. The effects are long lasting and withstand prolonged wear and washing.

Breathable fabrics

A layer of air trapped between the body and the fabric by the close, dense weave retains body heat in cold conditions and forms a cool insulation layer against outside temperatures in hot climates. Tiny pores in the fabric allow water vapour to pass out. This is called **wicking** and helps to regulate body heat.

Other microfibres include:
- Micromattique – often used for household fabrics
- Diolen Micro – superfine and often blended with wool, cotton or linen
- Meryl Micro – used for active sportswear
- Clarino and Sofrina – produced as alternatives to leather and used for fashion garments and luggage

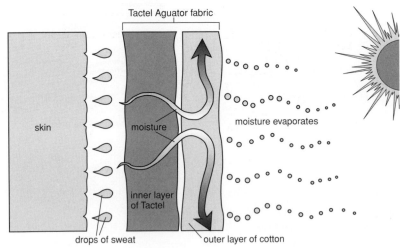

Construction of a microfibre fabric

Fibre blends

Blending fibres together gives the advantage of both, or all, fibres used to the fabric produced.

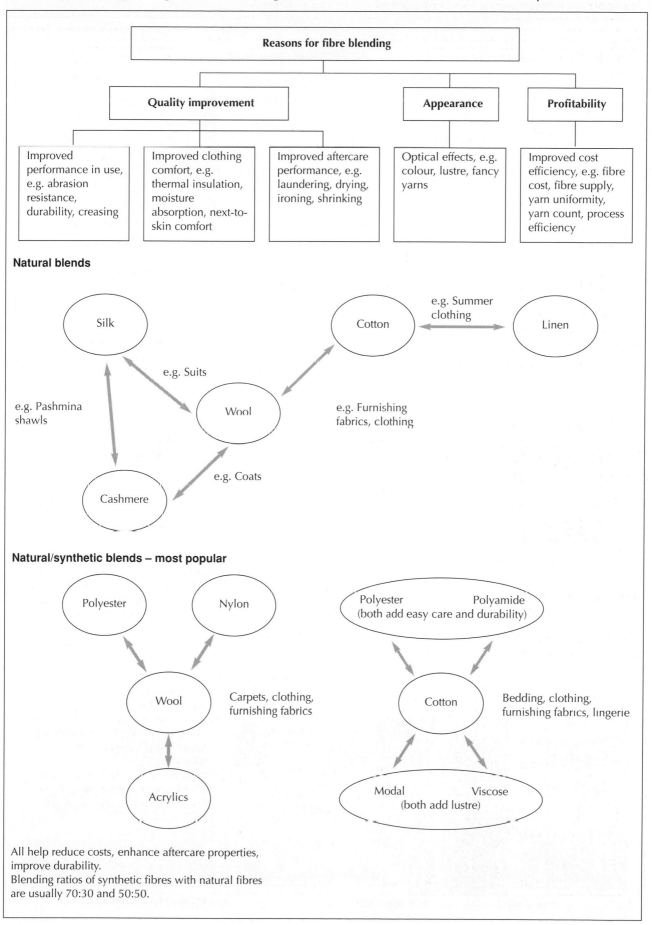

Reasons for fibre blending

Quality improvement

Improved performance in use, e.g. abrasion resistance, durability, creasing

Improved clothing comfort, e.g. thermal insulation, moisture absorption, next-to-skin comfort

Improved aftercare performance, e.g. laundering, drying, ironing, shrinking

Appearance

Optical effects, e.g. colour, lustre, fancy yarns

Profitability

Improved cost efficiency, e.g. fibre cost, fibre supply, yarn uniformity, yarn count, process efficiency

Natural blends

Silk

Cotton

e.g. Summer clothing

Linen

e.g. Suits

Wool

e.g. Pashmina shawls

e.g. Furnishing fabrics, clothing

e.g. Coats

Cashmere

Natural/synthetic blends – most popular

Polyester

Nylon

Wool

Carpets, clothing, furnishing fabrics

Acrylics

Polyester   Polyamide
(both add easy care and durability)

Cotton

Bedding, clothing, furnishing fabrics, lingerie

Modal   Viscose
(both add lustre)

All help reduce costs, enhance aftercare properties, improve durability.
Blending ratios of synthetic fibres with natural fibres are usually 70:30 and 50:50.

Spinning

TRADITIONAL SPINNING

Fibres are cleaned and combed into slivers

Slivers are either divided or folded before drawing or stretching out the fibres

Finally the fibres are twisted into a yarn

INDUSTRIAL SPINNING

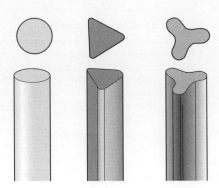

Spinneret shapes and fibre cross-sections

Final fibre produced depends on:
- the size and shape of the spinneret
- spinning and drawing conditions involved in the process

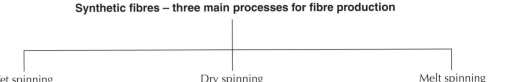

Synthetic fibres – three main processes for fibre production

Wet spinning e.g. viscose and acrylic	Dry spinning e.g. acrylic and acetate	Melt spinning e.g. nylon and polyester

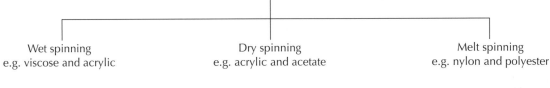

Spinning from a polymer solution		Spinning from a polymer melt
Wet spinning	Dry spinning	Melt spinning

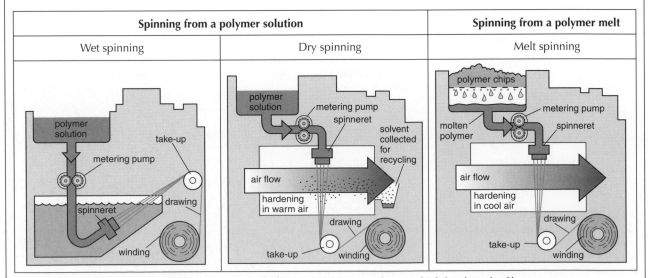

The molten solution is forced under pressure through the spinneret into a solution which hardens the fibres.

Yarn manufacture

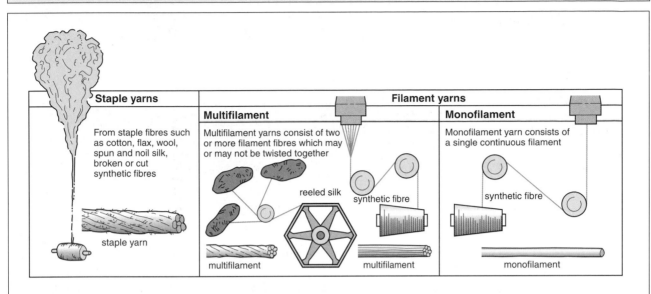

Staple yarns	Filament yarns	
	Multifilament	**Monofilament**
From staple fibres such as cotton, flax, wool, spun and noil silk, broken or cut synthetic fibres	Multifilament yarns consist of two or more filament fibres which may or may not be twisted together	Monofilament yarn consists of a single continuous filament

staple yarn

reeled silk   synthetic fibre   synthetic fibre

multifilament   multifilament   monofilament

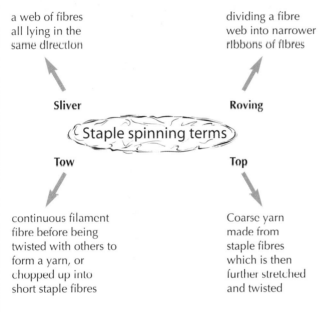

a web of fibres all lying in the same direction

dividing a fibre web into narrower ribbons of fibres

Sliver   **Roving**

Staple spinning terms

Tow   **Top**

continuous filament fibre before being twisted with others to form a yarn, or chopped up into short staple fibres

Coarse yarn made from staple fibres which is then further stretched and twisted

Yarn twist properties
Twist level = the number of twists per unit length

S twist   Z twist

← direction of twist →

COLOURING AND TEXTURING YARNS

Appearance	Process
Colour	Mixing different coloured dyed fibres during spinning
Structure	Altering the spinning process at regular or irregular intervals to produce long thicker sections of yarn between thin sections – a process called **slubbing**
Lustre	Using metal fibres or strips of metallized plastic film mixed with other fibres to produce Lurex-type yarns
Texture	Continuous filament yarns can be crimped and made bulky by a process which heats the yarn and then sets it into shape. Such bulk yarns are crimped to give them elasticity and looped for greater volume.

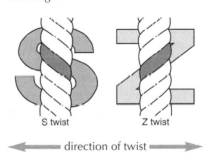

cotton   polyester

mixed yarn

Blended yarn

finer yarns   laminated metallic yarns

Lustre yarn

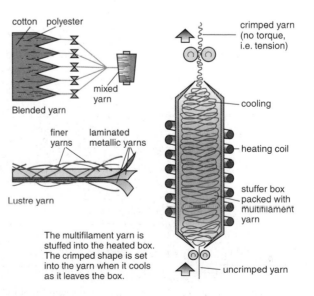

crimped yarn (no torque, i.e. tension)

cooling

heating coil

stuffer box packed with multifilament yarn

uncrimped yarn

The multifilament yarn is stuffed into the heated box. The crimped shape is set into the yarn when it cools as it leaves the box.

Weaving and knitting processes

Definitions
Weaving is a process where two yarns, the warp and weft, are woven together at right-angles to each other.
Knitting is a process of interlocking loops of yarn.

WEAVING – BASIC PRINCIPLES

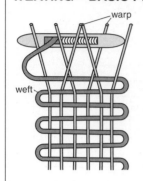

Shaft weaving involves the strong warp threads running the length of the loom with the weft threads being woven across using a shuttle. The way the warp and weft threads are woven produces different patterns or weaves.

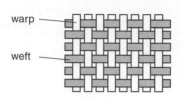

Plain weave   Satin weave   Twill weave

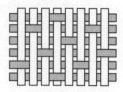

Jacquard weaving is an intricate weaving pattern against a plain background.

KNITTING – BASIC PRINCIPLES

Weft knitting is worked horizontally. The loops above and below each row interlock, holding the fabric together.

Weft knitting:

- unravels easily
- will ladder and run if cut or pulled
- is very stretchy and can lose shape very easily
- has a right and wrong side to the fabric

Warp knitting has warp yarns which interlock vertically along the length of the fabric from a separate yarn fed to each needle, in the same direction as the fabric grows.

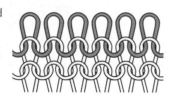

Warp knitting is an industrial process only and:

- is faster than weft knitting
- is cheaper to produce, with a wide variety of patterns
- only works well with filament yarn
- is less popular than weaving and weft knitting
- is difficult to unravel
- does not ladder or run when pulled or cut
- lies flat when cut
- has elasticity but keeps its shape well
- has identical right and wrong sides

Industrial knitting

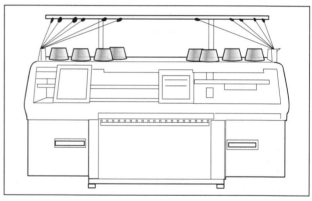

A flatbed knitting machine used to produce cardigans and jumpers, for example.

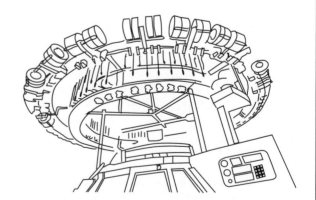

A circular knitting machine used to produce tubular products, for example, socks and tights.

Non-woven fabrics

Definition

A **non-woven fabric** is made directly from fibres without the production of yarn.

There are two main types of non-woven fabrics:
* felt (includes wool felts and needle felts)
* bonded

FELT

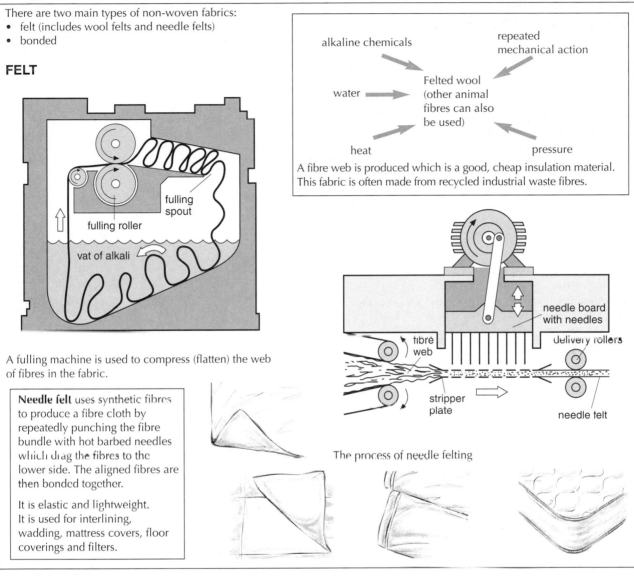

alkaline chemicals → Felted wool
(other animal
fibres can also
be used) ← repeated mechanical action

water →

heat → ← pressure

A fibre web is produced which is a good, cheap insulation material. This fabric is often made from recycled industrial waste fibres.

fulling spout
fulling roller
vat of alkali

A fulling machine is used to compress (flatten) the web of fibres in the fabric.

Needle felt uses synthetic fibres to produce a fibre cloth by repeatedly punching the fibre bundle with hot barbed needles which drag the fibres to the lower side. The aligned fibres are then bonded together.

It is elastic and lightweight. It is used for interlining, wadding, mattress covers, floor coverings and filters.

needle board with needles
delivery rollers
fibre web
stripper plate
needle felt

The process of needle felting

BONDED FABRICS

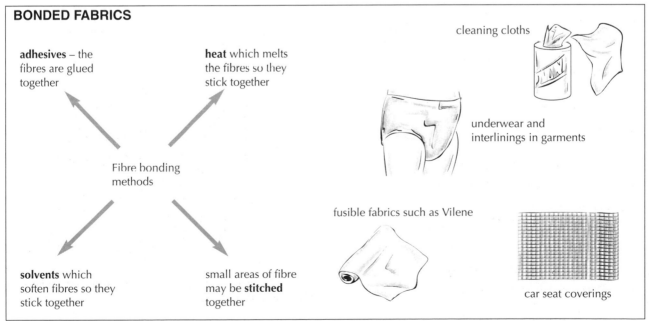

adhesives – the fibres are glued together

heat which melts the fibres so they stick together

Fibre bonding methods

solvents which soften fibres so they stick together

small areas of fibre may be **stitched** together

cleaning cloths

underwear and interlinings in garments

fusible fabrics such as Vilene

car seat coverings

Changing technologies

New non-woven fabrics are making a greater impact on the textile market because they are durable, flexible in use, easy to care for and are resistant to most chemicals.

Foams and rubber fabrics are thermoplastic – i.e. they can be moulded into shape. **Neoprene** is synthetic rubber mixed with stretch and knitted fabrics.

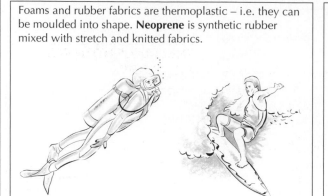

Lycra is a synthetic rubber fibre with elastane and is used for lingerie and in sportswear.

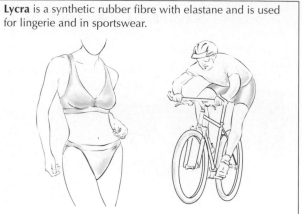

Fabrics from waste products such as **Polartec**, made from recycled plastic bottles. Totally biodegradable and used to line hand-knitted garments because it enhances performance.

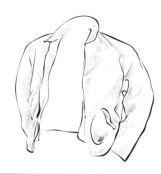

Tencel is produced from wood pulp. The fibre is made from lots of tiny fibrils and is used for all kinds of garments.

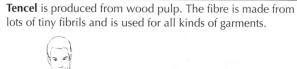

Composite fabrics such as **Kevlar** are used for very strong protective clothing.

Superabsorbent fibres such as **Oasis** are used for soaking up oils and other liquids.

Metallics – new metal yarns create fabrics as fluid as silk.

Laminated fabrics such as **Gore-Tex** are totally weatherproof but still allow the skin to breathe as they are semi-permeable.

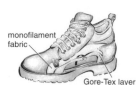

monofilament fabric

Gore-Tex layer

Preparation and intermediate fabric processing

Finishing processes are used to improve the final appearance, handle and wear of fabric. Products are seldom made from raw fabric: it is always dyed and/or printed before use. However, the use of raw-state fabrics such as calico is increasing due to environmental pressure to reduce chemical waste.

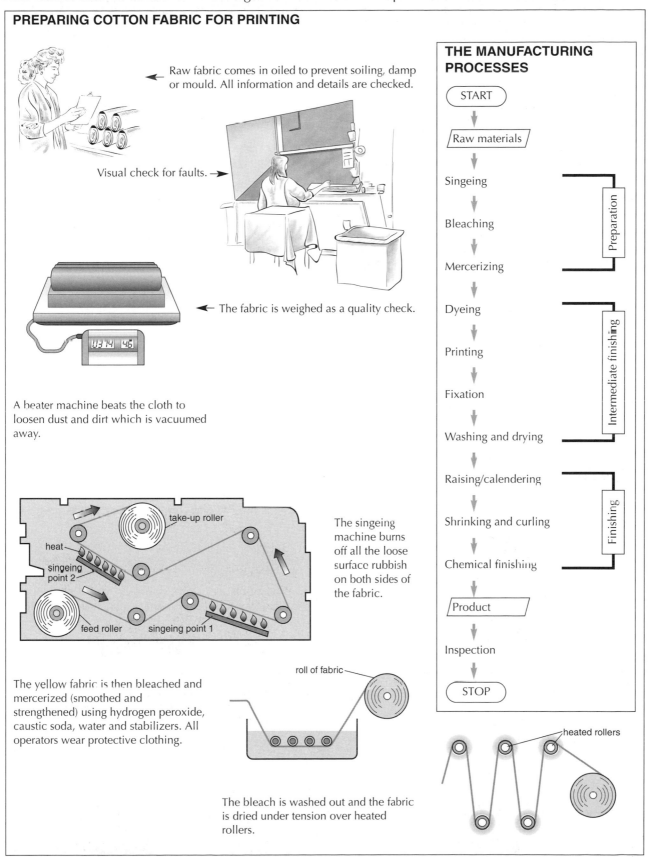

PREPARING COTTON FABRIC FOR PRINTING

← Raw fabric comes in oiled to prevent soiling, damp or mould. All information and details are checked.

Visual check for faults. →

← The fabric is weighed as a quality check.

A beater machine beats the cloth to loosen dust and dirt which is vacuumed away.

heat
take-up roller
singeing point 2
feed roller
singeing point 1

The singeing machine burns off all the loose surface rubbish on both sides of the fabric.

The yellow fabric is then bleached and mercerized (smoothed and strengthened) using hydrogen peroxide, caustic soda, water and stabilizers. All operators wear protective clothing.

roll of fabric

The bleach is washed out and the fabric is dried under tension over heated rollers.

heated rollers

THE MANUFACTURING PROCESSES

START
↓
Raw materials
↓
Singeing
↓
Bleaching   ⎤
↓   ⎬ Preparation
Mercerizing   ⎦
↓
Dyeing
↓
Printing   ⎤
↓   ⎬ Intermediate finishing
Fixation
↓
Washing and drying ⎦
↓
Raising/calendering ⎤
↓
Shrinking and curling ⎬ Finishing
↓
Chemical finishing ⎦
↓
Product
↓
Inspection
↓
STOP

Mechanical and chemical finishes

MECHANICAL FINISHES

Mechanical processes are dry processes used to create a finish on fabric.

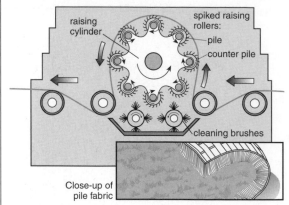

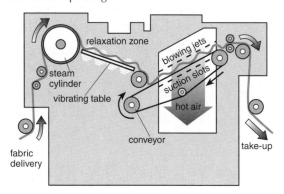

Raising: A series of spiked rollers is used to create a raised pile on fabric.

The **shrinking** process uses rollers and steam to shrink the fabric before printing.

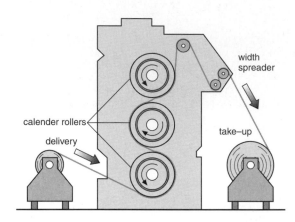

Calendering: The calender blows hot or cold air through the fabric, fluffing it up. The fabric is then rolled flat by the top roller. A cold roller gives a light, soft finish; a hot roller gives a harder, shiny finish but holds the colour better.

Different calendering effects are achieved by altering the surface of the rollers, changing their temperature, and adjusting their speed.

CHEMICAL FINISHING

Chemical finishing uses wet chemical solutions to change fabric properties or to enhance fabric performance.

RESISTANT

This label indicates that the fabric cannot be scorched or set on fire.

Scotchgard Fabric Protector

protected by [S]

• Resists staining   • Protection lasts
• Cleans easily   • Resists soiling

3M

After treating with Scotchgard, a fabric will be more resistant to dirt and stains and marks will clean off easily.

Finish	Process
Stain resistant	Applying specifically formulated substances which contain silicones or synthetic resins. Stain-resistant properties are applied mainly to fabrics used for clothing and carpeting.
Flame resistant	Applying particular flame resistant/retardant substances. Essential for fabrics used for furnishings, children's nightwear and protective clothing.
Water repellent	Spraying on specific chemicals such as silicones. Some clothing fabrics need to be water repellent, as do fabrics used for outdoor items such as tents.
Easy-care	Some frequently-used fabrics, such as cotton and viscose, need to be easy to care for. Chemicals are applied to protect the fabric against shrinkage and to make it resistant to creasing.
Anti-static	To make a fabric anti-static, the surface conductivity of its fibres needs to be improved. This helps to stop static charges building up on the fabric, important for floor coverings and synthetic fabrics used for clothing, especially underwear and lingerie.
Anti-pilling	Pilling is the appearance of small balls of fibre on the surface of a fabric. This often happens with woollen fabrics or those made from synthetic fibres. Anti-pilling properties are applied by using solvents or polymers that form a film on the fabric's surface.
Moth proof	Woollen fabrics often need protection from moths. Chemicals applied to the fabric repel moths by making the fibres inedible.

Dyeing and printing processes

Dyeing and printing are both wet processes in which the dye is dissolved or dispersed in water.

DYEING

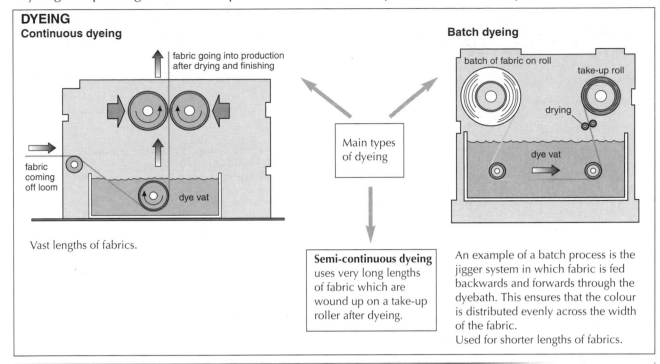

Continuous dyeing

fabric going into production after drying and finishing

fabric coming off loom

dye vat

Vast lengths of fabrics.

Batch dyeing

batch of fabric on roll

take-up roll

drying

dye vat

Main types of dyeing

Semi-continuous dyeing uses very long lengths of fabric which are wound up on a take-up roller after dyeing.

An example of a batch process is the jigger system in which fabric is fed backwards and forwards through the dyebath. This ensures that the colour is distributed evenly across the width of the fabric.
Used for shorter lengths of fabrics.

PRINTING

Screen printing uses very expensive stencil screens to print a pattern on fabric.

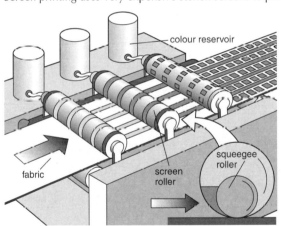

colour reservoir

squeegee roller

fabric

screen roller

Rotary screen printing
- CAD (computer-aided design) systems are used to develop the cylindrical screens.
- The pattern design is built up as the fabric passes underneath each cylindrical screen.
- Each screen costs about £300 to produce and some patterns use 15 screens to build the design.

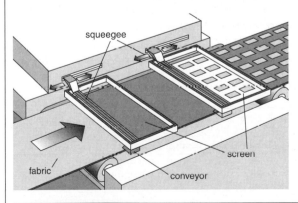

squeegee

fabric

screen

conveyor

This simple flower design was made up and printed on a pale lilac background as follows (use your imagination!):
1 white flower shapes
2 green leaf shapes
3 blue flower shapes
4 yellow centres of flowers
5 navy wavy line, flower and leaf outlines
6 orange stipple effect on flower centres

Flat screen printing
- The fabric is held flat and secure on the conveyor belt.
- One screen width is printed at a time, the fabric being moved one screen width each time.
- Accounts for 18% of commercial printing.

Modern **inkjet printing** techniques are now being increasingly used for fabric printing.

Fabric aftercare

There are four factors in washing and cleaning textile products:

- **Temperature** (washing, ironing and drying temperatures)
- **Mechanical action** (agitation to loosen dirt from the fibres)
- **Washing medium** (detergent dissolved in water)
- **Time** (soaking, washing, drying times)

International care labelling code (ITCLC)

Care	Symbol	Meaning
Washing		Hand or machine washing is acceptable.
		The number in the 'wash tub' indicates the maximum water temperature the fabric and/or product should be washed at.
		Use a synthetic wash.
		Use a wool wash.
		Hand wash only.
		Do not wash.
Drying		Drying symbols tell you if the product should be drip dried, line dried or dried flat to maintain its shape.
Tumble drying		Product can be tumble dried safely. Sometimes temperature and time for drying are indicated. Most textile items can be safely tumble dried on low heat settings. Synthetics must never be dried on high heat. Tumble driers should not be overloaded.
Ironing		The more dots, the higher the temperature. Modern irons also carry these symbols on the control panel to help people choose the correct setting.
Dry cleaning	All solvents (A) (P) Perchloroethylene (F) Certain solvents only	These symbols are important because they inform the dry cleaners how to clean the product. The letters indicate the type of cleaning solvent it is safe to use on the fabric.
Chlorine	△a	Chlorine bleach can be used.

Fabric aftercare labelling

The code and fabrics

	Cotton	Linen	Wool	Silk	Viscose/Modal	Acrylic	Polyester	Nylon
Washing	95° White / 60° Colours / 40° Dark colours	95° White / 60° Colours / 40° Dark colours	30°	30°	40° Viscose / 60° Modal	40°	50°	40°
Chlorine	△a	△a	✕	✕	✕	✕	✕	✕
Ironing	⚫⚫⚫	⚫⚫⚫	⚫⚫	⚫	⚫⚫	⚫	⚫⚫	⚫ Without steam
Dry cleaning	Ⓐ	Ⓐ	Ⓟ	Ⓟ	Ⓟ	Ⓟ	Ⓟ	Ⓟ
Drying	⊙⊙	⊙⊙	✕	✕	✕ Viscose / ⊙ Modal	✕	⊙	✕

Stain	Removal method
Protein-based stains (blood, egg, perspiration)	Wash in biological powder
Drinks (tea and coffee)	Wash in biological powder
Ballpoint pen	Dab with methylated spirits, then wash
Emulsion paint	Rinse straight away in cold water
Oil-based paint	Dab with white spirit and wash
Chewing gum	Place the item in the freezer and pick off the gum when it is frozen

Golden rules of stain removal

- Remove the stain as early as possible.
- Rinse the area in cold water (hot water 'sets' stains).
- Soak the stain in biological or enzyme washing powder solutions so that the enzymes can break down the stain.
- Always check the label for colourfastness.
- Use specialist solvents with care, read the directions carefully.

Research and gathering information

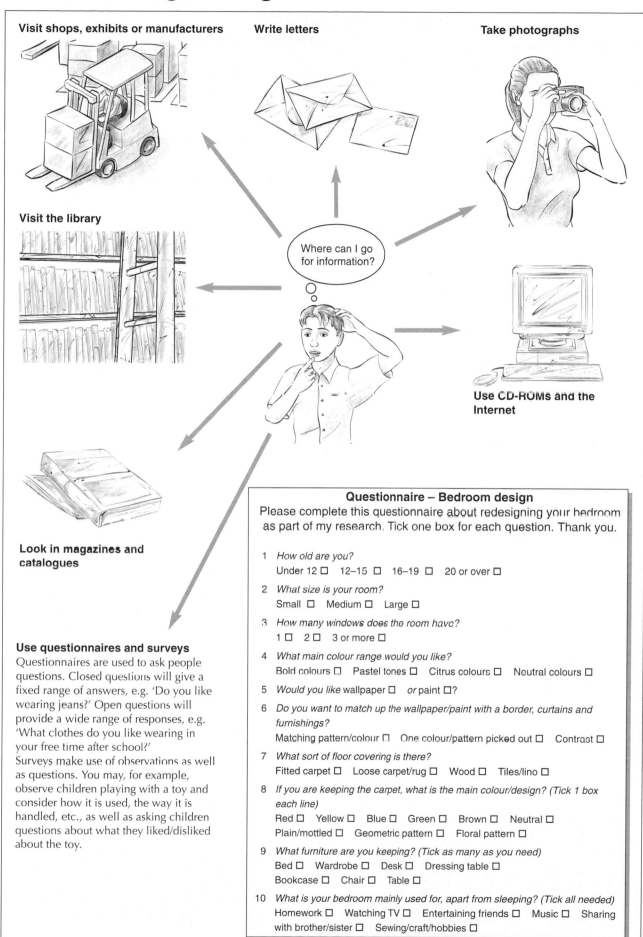

Visit shops, exhibits or manufacturers

Write letters

Take photographs

Where can I go for information?

Visit the library

Use CD-ROMs and the Internet

Look in magazines and catalogues

Use questionnaires and surveys

Questionnaires are used to ask people questions. Closed questions will give a fixed range of answers, e.g. 'Do you like wearing jeans?' Open questions will provide a wide range of responses, e.g. 'What clothes do you like wearing in your free time after school?'

Surveys make use of observations as well as questions. You may, for example, observe children playing with a toy and consider how it is used, the way it is handled, etc., as well as asking children questions about what they liked/disliked about the toy.

Questionnaire – Bedroom design

Please complete this questionnaire about redesigning your bedroom as part of my research. Tick one box for each question. Thank you.

1 *How old are you?*
   Under 12 ☐   12–15 ☐   16–19 ☐   20 or over ☐

2 *What size is your room?*
   Small ☐   Medium ☐   Large ☐

3 *How many windows does the room have?*
   1 ☐   2 ☐   3 or more ☐

4 *What main colour range would you like?*
   Bold colours ☐   Pastel tones ☐   Citrus colours ☐   Neutral colours ☐

5 *Would you like* wallpaper ☐   *or* paint ☐?

6 *Do you want to match up the wallpaper/paint with a border, curtains and furnishings?*
   Matching pattern/colour ☐   One colour/pattern picked out ☐   Contrast ☐

7 *What sort of floor covering is there?*
   Fitted carpet ☐   Loose carpet/rug ☐   Wood ☐   Tiles/lino ☐

8 *If you are keeping the carpet, what is the main colour/design? (Tick 1 box each line)*
   Red ☐   Yellow ☐   Blue ☐   Green ☐   Brown ☐   Neutral ☐
   Plain/mottled ☐   Geometric pattern ☐   Floral pattern ☐

9 *What furniture are you keeping? (Tick as many as you need)*
   Bed ☐   Wardrobe ☐   Desk ☐   Dressing table ☐
   Bookcase ☐   Chair ☐   Table ☐

10 *What is your bedroom mainly used for, apart from sleeping? (Tick all needed)*
   Homework ☐   Watching TV ☐   Entertaining friends ☐   Music ☐   Sharing with brother/sister ☐   Sewing/craft/hobbies ☐

Product evaluation

Look for: available textile products that fulfil some or all of the functions that you want your product to perform.
Even look at products made from other materials and for other purposes to gain ideas.

Look for:

possible conflicts in the design and how they have been resolved

how improvements have been used on a basic design idea

the main features used and how you can measure or test performance

the quality of the item versus its cost

how well it meets a need within the cost

elastic

Compare products using a **spreadsheet** format or a **star diagram**.

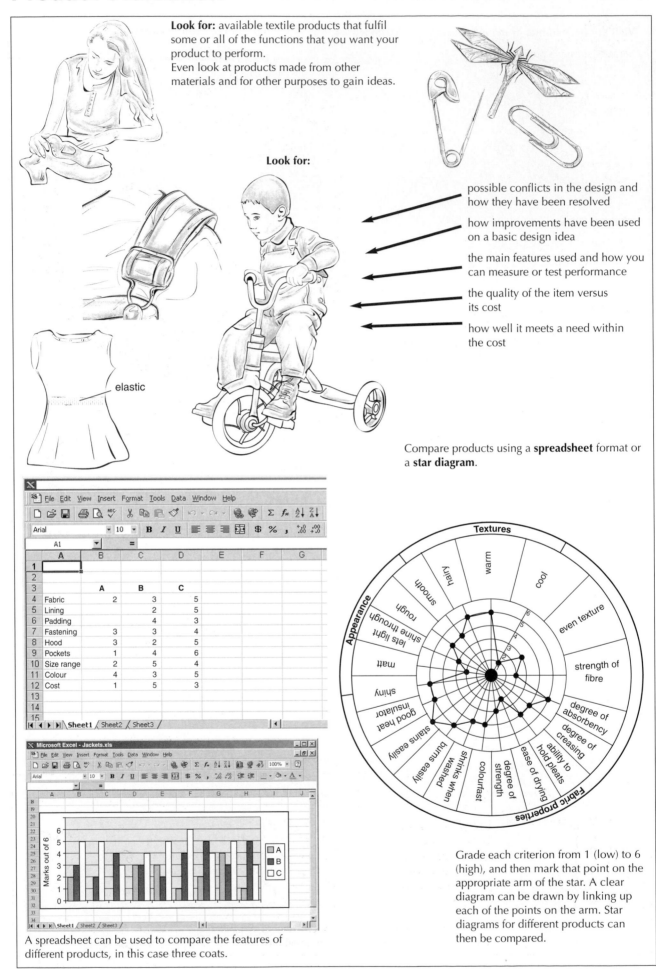

A spreadsheet can be used to compare the features of different products, in this case three coats.

Grade each criterion from 1 (low) to 6 (high), and then mark that point on the appropriate arm of the star. A clear diagram can be drawn by linking up each of the points on the arm. Star diagrams for different products can then be compared.

Colour, line and shape

COLOUR

Colour draws attention to particular features and characteristics of a product. It can:

- affect mood – sunny yellow

- create warmth – red, pink, yellow – or coolness – blue or lilac

- create excitement – jazzy colours – or peace – white

Find a colour version of the wheel!

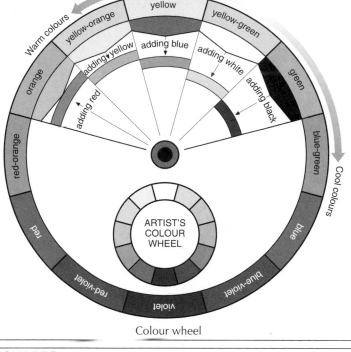

Colour wheel

Using colours:

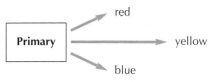

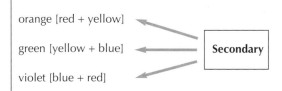

	six colours produced by mixing a primary and a secondary colour together
Tertiary	

Harmonious	colours lying side by side on the colour wheel

Contrast	colours opposite each other on the colour wheel

Tone	adding black or white changes the tone of a colour

SHAPES

Geometric shapes are useful as a basic tool for shape development.

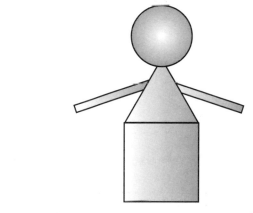

LINES

Lines organize visual form. The line is a basic design tool used to create styles, shapes, patterns, forms and textures.

General rules are:

Vertical lines

Adds height and also narrows a profile.

Horizontal lines

Shortens and adds width to a profile.

Diagonal lines

Adds a feeling of movement and instability.

Curved lines

Softens and emphasizes natural curves.

Mood and story boards

Definitions
A **mood board** describes your thoughts about what you want to achieve in a product, for example colour, pattern, main features.
A **story board** is a series of sketches, diagrams or photographs to show the main stages of manufacture.

EXAMPLES OF MOOD BOARDS

Summer theme

Sport theme

You could use a mixture of resources to make a mood board, with photographs and pictures from magazines, swatches of colour, fabric or patterns. You could also add trims and buttons or draw diagrams and sketches.

A STORY BOARD FOR MAKING A BAG

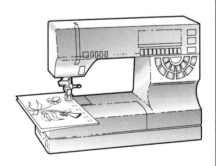

1 Print the fabric

2 Cut out the pattern pieces

3 Sew the side seams

4 Make the casing at the top

5 Thread the draw string through

6 Make and attach the handle

Spatial relationships

Definition
Spatial relationships means the way we see products in relation to the other things around them.

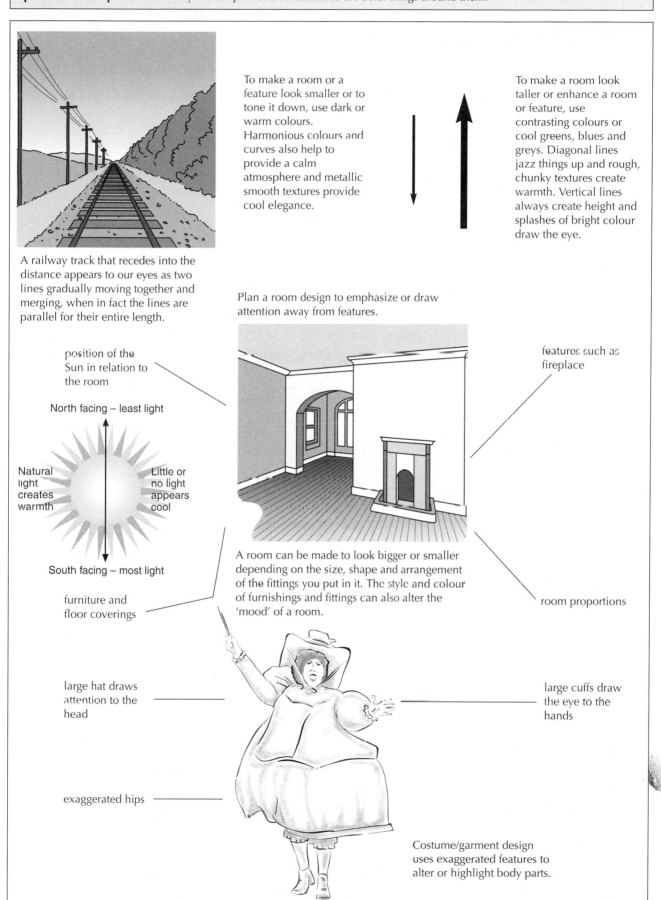

To make a room or a feature look smaller or to tone it down, use dark or warm colours. Harmonious colours and curves also help to provide a calm atmosphere and metallic smooth textures provide cool elegance.

To make a room look taller or enhance a room or feature, use contrasting colours or cool greens, blues and greys. Diagonal lines jazz things up and rough, chunky textures create warmth. Vertical lines always create height and splashes of bright colour draw the eye.

A railway track that recedes into the distance appears to our eyes as two lines gradually moving together and merging, when in fact the lines are parallel for their entire length.

Plan a room design to emphasize or draw attention away from features.

position of the Sun in relation to the room

North facing – least light

Natural light creates warmth

Little or no light appears cool

South facing – most light

furniture and floor coverings

features such as fireplace

room proportions

A room can be made to look bigger or smaller depending on the size, shape and arrangement of the fittings you put in it. The style and colour of furnishings and fittings can also alter the 'mood' of a room.

large hat draws attention to the head

large cuffs draw the eye to the hands

exaggerated hips

Costume/garment design uses exaggerated features to alter or highlight body parts.

Drawing body proportions

When developing any garment design it is important to know how to draw silhouettes in proportion and in different poses.

eye
nose
mouth
shoulder
bust
waist
hip
knee
ankle

You can use a line of head shapes as the basis for drawing figures. These are the normal proportions for men and women (above) and teenagers and children (below).

Teenager 9/10 years   5 years   2 years

Annotation, labels and clear, brief notes can provide any other information not clear in your sketches.

Notes should not cover any part of the drawing.

Add detail to your drawings to complete the picture:

- Reflect the style of the garment, light for dresses, heavy for winter coats.
- Include swatches of fabric on the side of the drawing.
- Use a fine black pen to highlight important details.
- Use coloured marker pens to give colour and tone to the garment shape.
- Experiment with paints, pens, etc. to create a patterned or textured appearance.

Textile components

Textile components can be functional and/or decorative.

COMPONENT	APPLICATION	ADVANTAGES
Covered buttons (buttons covered with product fabric)	Popular for cushions, curtains and bedlinen. Garments that are made from printed fabric can have covered buttons.	Can provide an interesting contrast, or blend in with background.
Stud fastenings	Popular in denim jackets and jeans, waistbands, bag fastenings (often hidden under a mock buckle).	Quick and easy to use.
Eyelets	Useful where a lacing effect is required – front of garment, sleeve or neckline; top of bag.	Very decorative; firm when closed, roomy when opened.
Ribbon	Stitched into position or woven through finished holes, to gather fabric or tie edges together. Wide ribbon is useful for creating a loop effect at the top of curtains.	Range of widths and colours; strong in use and can have decorative designs printed on it. Can be a harmonious or contrasting feature.
Lace	As a decorative feature at neck, sleeve, pocket or hem edge, in panels or as a frill edge to a product.	Range of colours, styles and widths available; very decorative and feminine; can be harmonious or contrasting.
Braid	Added decoration around the neck, sleeve and hem edges; decorative feature for products, lampshades, curtains, cushions, upholstery.	Various widths and colours; can provide contrasting features to highlight and enhance a product.
Buckles	Used decoratively, as well as on belts; can be metal or covered in product fabric. Front buckles can attach straps to main part of the product.	Helps to emphasize a part of a garment or product such as a hipline or a handle.
Chains	Chains can used very decoratively on garments to emphasize specific features, or to perform functions, e.g. a bag shoulder strap.	Cannot be slashed by bag snatchers.
Safety pins	Popular feature; holding sections together or just for decoration.	Can hold garment sections together in a decorative way.
Rivets	Used for fastenings and reinforcement, particularly on denim items.	Strengthens strain points; rivetted buttons are stronger than stitched ones.
Toggles	Used at the end of a drawstring or as a fastening. Some have spring grips to hold the cord under tension so that it does not have to be tied.	Very interesting shapes and sizes available. Why not make your own?
Iron or sew-on embroidered logos or motifs	Traditionally used for school badges but expanded to a wide variety of designs; excellent for covering worn areas of garments; also decorative.	Helps to individualize products. Very easy to apply.
Zips	Used for closures; ideal for close-fitting garments.	Can be concealed or conspicuous.

Equipment – small

Small equipment should always be kept well organized and maintained.

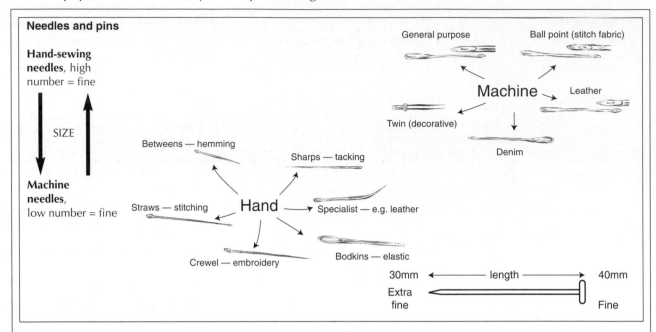

Needles and pins

Hand-sewing needles, high number = fine

SIZE

Machine needles, low number = fine

Betweens — hemming

Sharps — tacking

Straws — stitching

Hand

Specialist — e.g. leather

Crewel — embroidery

Bodkins — elastic

General purpose

Ball point (stitch fabric)

Machine

Leather

Twin (decorative)

Denim

30mm ← length → 40mm

Extra fine

Fine

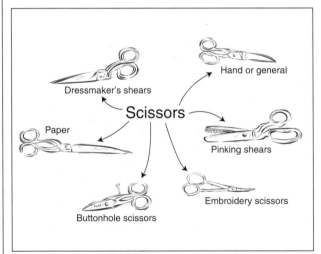

Dressmaker's shears

Hand or general

Scissors

Paper

Pinking shears

Buttonhole scissors

Embroidery scissors

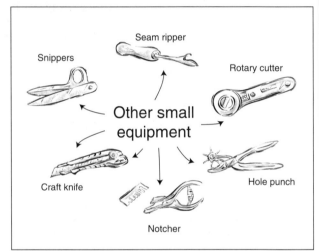

Snippers

Seam ripper

Rotary cutter

Other small equipment

Craft knife

Hole punch

Notcher

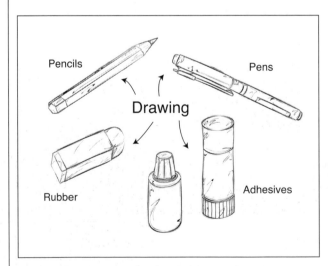

Pencils

Pens

Drawing

Rubber

Adhesives

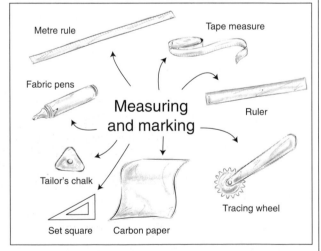

Metre rule

Tape measure

Fabric pens

Measuring and marking

Ruler

Tailor's chalk

Tracing wheel

Set square

Carbon paper

Threads
- hand threads
- machine threads

Choose a thread one shade darker than the fabric and with similar fibre content if possible; cotton for natural fabrics and polyester for silk and synthetic fabrics.

Equipment – large

Most items of large equipment are electrically operated and many are computer controlled. The item you will use most is a sewing machine so make sure that you practise to become confident in the use of a range of machines. Remember that sticking to just one not only may restrict the techniques available to you, but could cause you problems if it breaks down.

SEWING MACHINES

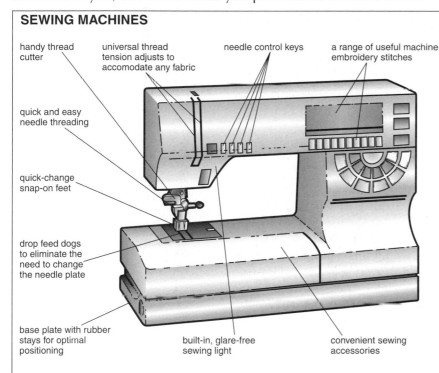

handy thread cutter

universal thread tension adjusts to accomodate any fabric

needle control keys

a range of useful machine embroidery stitches

quick and easy needle threading

quick-change snap-on feet

drop feed dogs to eliminate the need to change the needle plate

base plate with rubber stays for optimal positioning

built-in, glare-free sewing light

convenient sewing accessories

Optional feet can provide versatility in your work, e.g.

- appliqué foot – for narrow satin stitch for appliqué and logos
- cording foot – for easy-to-sew pin tucks
- buttonhole foot – for making buttonholes
- felling foot – automatically folds the fabric edges for a machine fell or double-stitched seam
- edge-hemming foot – gives narrow double hem

Typical features on a sewing machine

Other machines you may use

An overlocking machine for seams

A knitting machine (can be computer controlled)

A POEM embroidery machine (computer controlled)

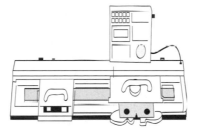

A CAMM 1 machine is a computer-aided cutter plotter useful for making stencils.

PRESSING EQUIPMENT

Item	Use
Steam iron	Water is stored in a reservoir and heated by electric elements to form steam which is released through small holes in the base of the iron. The temperature of the iron must be carefully regulated to suit the type of fabric.
Ironing board	Used for pressing large areas. Special covers make more efficient use of the heat from the iron.
Sleeve board	Used to press tubular pieces such as sleeves.
Ironing cushion	Helps when pressing difficult shapes such as shoulders on garments (also known as a 'ham' because of its shape).

Hazards in the workplace

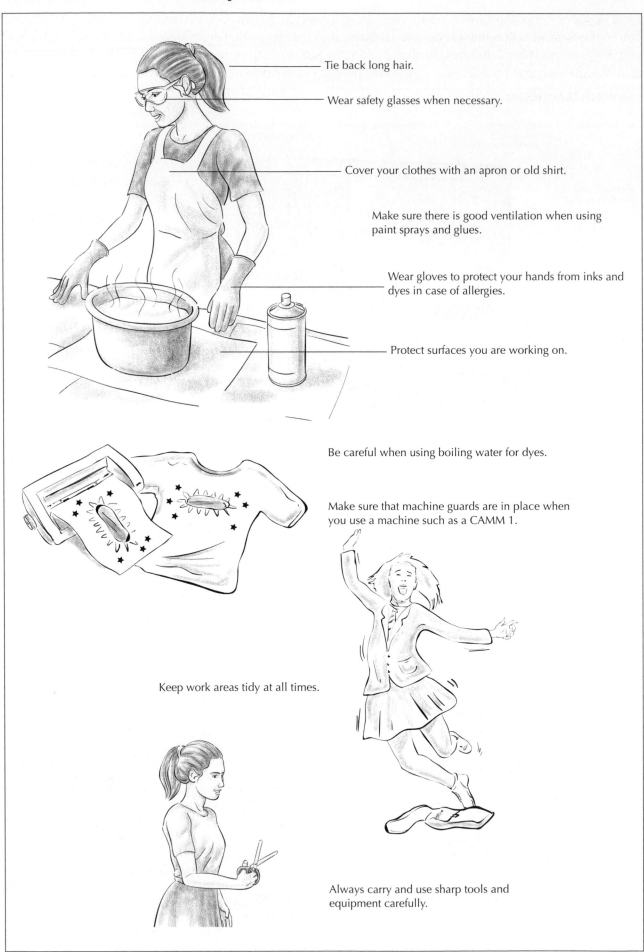

Tie back long hair.

Wear safety glasses when necessary.

Cover your clothes with an apron or old shirt.

Make sure there is good ventilation when using paint sprays and glues.

Wear gloves to protect your hands from inks and dyes in case of allergies.

Protect surfaces you are working on.

Be careful when using boiling water for dyes.

Make sure that machine guards are in place when you use a machine such as a CAMM 1.

Keep work areas tidy at all times.

Always carry and use sharp tools and equipment carefully.

Hazards in products

Definitions
British Standards Institution (BSI): sets the safety standards for a wide range of products.
Kitemark: products given the kitemark logo are tested regularly against BSI standards.
British Electrotechnical Approval Board (BEAB): tests electrically powered products.

FIBRES AND FABRICS

Will the fabric burn or melt? Flame resistance needed for nightclothes, wadding for toys, protective clothing, furnishings.

Children's clothes and toys must be flame resistant.

Can fibres be pulled out and become lodged in the throat of a child?

Toys must be safe for young children to handle.

TEMPERATURE

Can items withstand extremes of heat or cold? e.g. oven gloves, sun shade, clothing for Arctic weather.

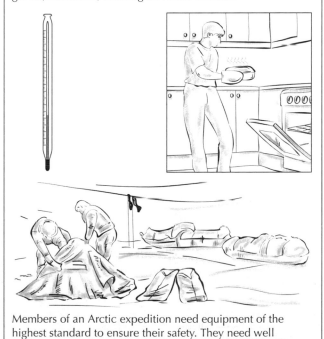

Members of an Arctic expedition need equipment of the highest standard to ensure their safety. They need well designed and properly insulated clothing and sleeping bags, windproof tents, and bags made of rough, waterproof rip-stop fabric for their supplies.

FASTENINGS

Can the fastenings used:
- be swallowed?
- trap a small finger?
- be used easily by the elderly or disabled?
- get trapped in a door?

Some features of clothing can be unexpectedly hazardous.

SHARP EDGES

Does the fabric have to cover sharp corners or edges? Is there an internal wire to protect? Is the product being designed to hold a sharp tool such as scissors?

The British kitemark identifies products that have passed rigorous safety checks.

Quality

Looking for quality in a product

A high-quality textile product is one that is reliable in use over a long period of time.

Product quality depends on:
- design features
- fabrics used
- assembly processes
- finishing techniques
- quality control checks

Poor quality has **hidden costs**:
- high wastage
- assembly problems
- frequent repairs
- customer dissatisfaction

Quality of design

- zip
- ease of use of handles
- additional features such as side pockets
- sits flat on the ground

Quality of manufacture

- handle
- durable fabric used
- good quality components used
- strong stitching

Quality assurance is the way the production system is managed to make sure that the specifications for the product are met.

Quality control is the set of tests and inspections applied at specific points in the production process.

1 Check fabrics and components	→	2 Check correct cutting and bundling	→	3 Check machining and seam finishing

6 Final inspection check	←	5 Check handle alignment and assembly	←	4 Check zip insertion

Quality assurance and manufacturing cost considerations:

✔
- Monitoring systems to maintain standards
- Tests/check cost
- Extra labour

✖
- Faulty goods returned
- Reworking labour
- Repairing labour
- Extra use of energy
- Fabric/component waste
- Loss of customer support

Critical control points

These are the points in production when the fabric, components or product being assembled must be checked to ensure quality standards are met.

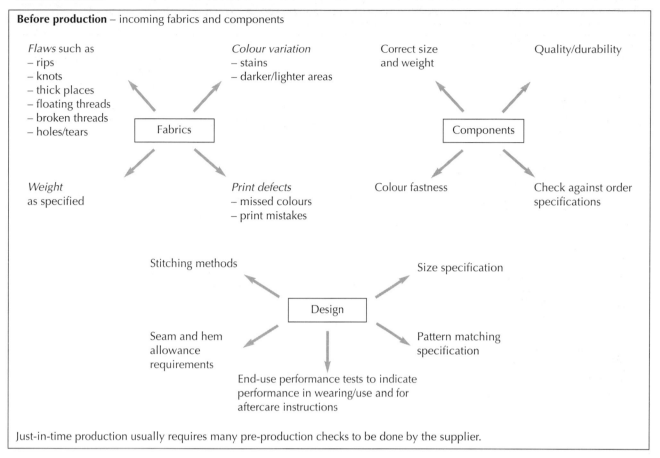

Before production – incoming fabrics and components

Flaws such as
– rips
– knots
– thick places
– floating threads
– broken threads
– holes/tears

Colour variation
– stains
– darker/lighter areas

Correct size
and weight

Quality/durability

Fabrics

Components

Weight
as specified

Print defects
– missed colours
– print mistakes

Colour fastness

Check against order
specifications

Stitching methods

Size specification

Design

Seam and hem
allowance
requirements

Pattern matching
specification

End-use performance tests to indicate
performance in wearing/use and for
aftercare instructions

Just-in-time production usually requires many pre-production checks to be done by the supplier.

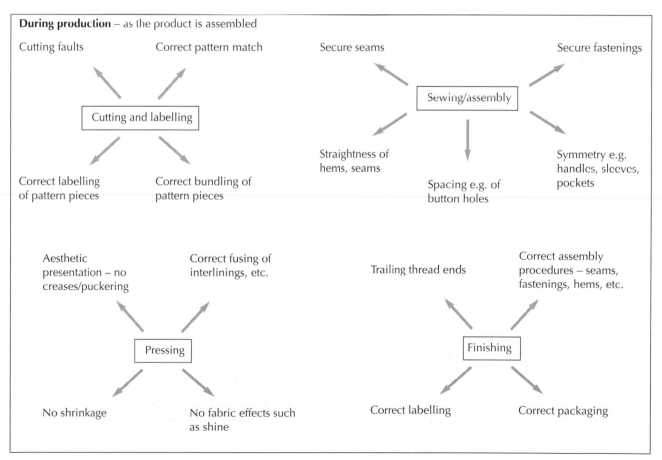

During production – as the product is assembled

Cutting faults

Correct pattern match

Secure seams

Secure fastenings

Cutting and labelling

Sewing/assembly

Correct labelling
of pattern pieces

Correct bundling of
pattern pieces

Straightness of
hems, seams

Spacing e.g. of
button holes

Symmetry e.g.
handles, sleeves,
pockets

Aesthetic
presentation – no
creases/puckering

Correct fusing of
interlinings, etc.

Trailing thread ends

Correct assembly
procedures – seams,
fastenings, hems, etc.

Pressing

Finishing

No shrinkage

No fabric effects such
as shine

Correct labelling

Correct packaging

Timing and planning

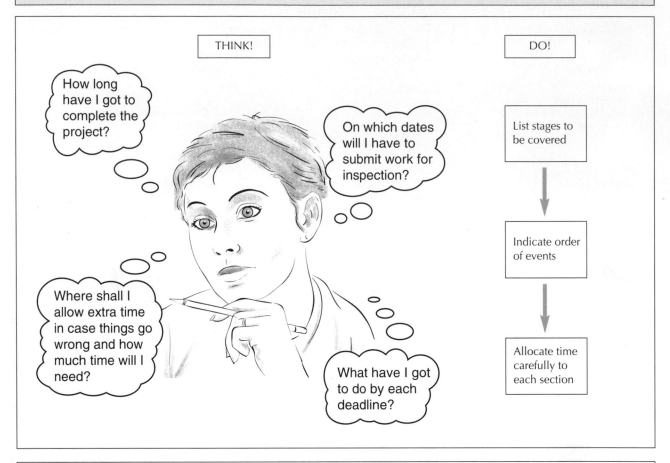

THINK!

How long have I got to complete the project?

On which dates will I have to submit work for inspection?

Where shall I allow extra time in case things go wrong and how much time will I need?

What have I got to do by each deadline?

DO!

List stages to be covered

↓

Indicate order of events

↓

Allocate time carefully to each section

ORGANIZE AND PLAN

What information do I need to gather?

For example:
- information about customers
- available products for evaluation
- properties of fabrics
- different processing methods
- information about what the product needs to be able to do

How can I get this information?
- printed materials
- computer materials
- questionnaires
- surveys
- disassembling products
- evaluation of products

How much time can I allow to:
- do my research
- develop my design ideas
- investigate production processes
- make a prototype product
- test and evaluate

What do I need to measure?

For example:
- body size
- room size
- furniture size
- weights to be carried/stored
- size and shapes of items to be stored
- what features the customer wants
- what price the customer will pay
- the strength of the fabric
- how much light each fabric will allow through

What equipment and resources do I need?

For example:
- large sewing equipment
- small sewing equipment
- fabric
- components
- other materials
- fabric printing equipment
- testing equipment
- pressing equipment

Producing a prototype

Definition
A **toile** is a model of a garment made out of calico or other cheap fabric.

It is always wise to make a simple prototype model of a garment or any other textile product out of cheap fabric. This allows you to identify property details and measurements before you make your final fabric choice for the product. Prototypes can also be produced to test specific parts of a product, for example, an adjustable strap.

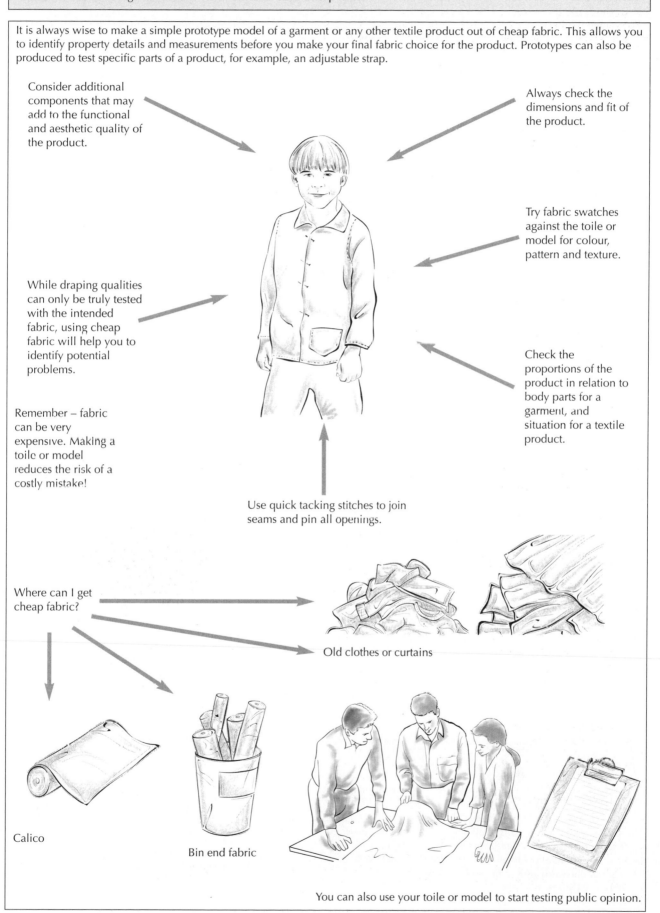

Consider additional components that may add to the functional and aesthetic quality of the product.

Always check the dimensions and fit of the product.

Try fabric swatches against the toile or model for colour, pattern and texture.

While draping qualities can only be truly tested with the intended fabric, using cheap fabric will help you to identify potential problems.

Check the proportions of the product in relation to body parts for a garment, and situation for a textile product.

Remember – fabric can be very expensive. Making a toile or model reduces the risk of a costly mistake!

Use quick tacking stitches to join seams and pin all openings.

Where can I get cheap fabric?

Old clothes or curtains

Calico

Bin end fabric

You can also use your toile or model to start testing public opinion.

Joining

TACKING

Tacking is used to mark fitting lines and alterations on single fabric and for holding layers of fabric in position for their final stitching.

Tacking

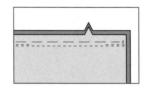

Basting or diagonal tacking

SEAMS
Choice of seam depends on the fabric being used, the type and purpose of the product, and the position of the seam.

A plain seam

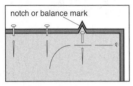
notch or balance mark

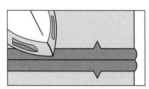

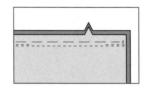

1 Place the right sides of the fabric together, being sure to match the fitting lines and balance marks. Pin and tack the fabric into position.

2 Remove the pins and stitch along the fitting lines. Remove the tacks and press along the stitching.

3 Press the turnings open ready for the edges to be finished (see below).

Methods of finishing the seam edges of a plain seam

Pinked and stitched

Bias-bound

Turned and stitched

Edge stitched

French seam
A French seam is good for products made from shear fabrics and infant clothes because it is inconspicuous and strong but not bulky.

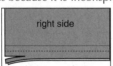
right side

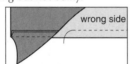
wrong side

Place the pieces of fabric *wrong* sides together, and stitch a plain seam 1 cm from the edge. Trim the seam allowance to 3 mm.

Turn the garment inside out (right sides together), then press the fabric flat at the seam line, with stitched line exactly on edge of fold. Stitch on seam line 4 mm from the edge.

Flat-felled seam
A flat-felled seam is strong and decorative.

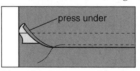
press under

Place fabric wrong sides together and stitch on the seam line. Press seam open, and then to one side. Trim inner seam allowance to 5 mm. Press the edge of the outer allowance under. Stitch this folded edge to the garment, keeping distance between the lines of stitching even.

Double-stitched seam
Double-stitched seams are strong, often used for stretch fabrics.

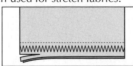

Place pieces of fabric right sides together and stitch a plain seam. Stitch a second row 3 mm from the first. Trim seam allowances. Press seam to one side. Useful for jersey fabrics, which tend to roll.

A zigzag stitch can be used in place of the second line of plain stitching – useful for fabrics which fray a lot, or for loosely knitted jerseys.

Overlocked seam
Overlocking is an industrial method. It is strong and very quick to do.

With an overlocking machine you can sew, trim and neaten a seam all at the same time.

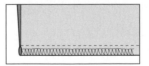

Shaping

Definitions

Darts control excess fabric, moulding it to fit the shape of the body by the use of folds that end in a point at the fullest part of the shape.

Tucks control fullness and ease of movement by the use of small folds of fabric held by machine stitching.

Gathering and the use of **elastication** are both methods of gathering the fabric up into evenly spaced gathers.

DARTS

Single-pointed dart

1 Fold fabric right sides together with fitting lines matching, then pin and tack along the line.

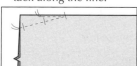

2 Start at the wide end, machine stitch along the fitting line until the fold is reached, then reverse and add three stitches along the fold. Remove tacks and press.

Double-pointed dart

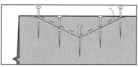

1 Fold fabric right sides together, matching fitting lines, and tack.

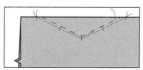

2 Machine stitch along the fitting line out of the centre of the dart, and up to the top point. Reverse machine stitching to secure.

There are three main types of dart:
- Single-pointed dart
- Double-pointed or contour dart
- Dart tuck (open-ended dart)

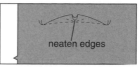

neaten edges

3 Remove tacks and press. To prevent dragging on the right side of the dart, it should be snipped on the wrong side at the centre of the dart to within 3 mm of the machine stitching. Neaten raw edges with loop stitch.

TUCKS

Dart tuck

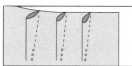

Fold fabric right sides together with fitting lines matching. Pin and tack. Starting from side end, machine stitch along the fitting line to end of tuck, then reverse machine stitching for 1.25 cm to finish and strengthen the stitching. Remove tacking. Press stitching, then press tuck to required position.

Pin tuck

Pin tucks can be highly decorative and are often used on blouses and evening shirts.

Pressing darts and tucks

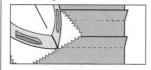

Press each tuck flat as it was stitched. Use a cloth to press darts and tucks flat to avoid marking the fabric. When pressing darts on garments: press waist darts to centre front and centre back, underarm darts towards the waistline, shoulder darts towards the neck and elbow darts towards the wrist.

GATHERING

Gathering draws in the fullness of a garment evenly. Modern sewing machines make gathering easy by setting the stitch length to the longest.

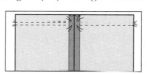

Use a loose tension and the longest stitch available. Work two rows of stitching with the first row exactly in line with the second. Do not stitch across a seam; gather each section separately.

ELASTICATION

Elastication provides a casing through which elastic may be passed. The casing can be in two forms:
- a fabric strip is sewn to the inside of the fabric edge
- the fabric edge is turned over

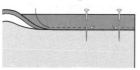 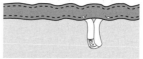

1 Measure and press a 0.5 cm turning. Fold a second turning down, wide enough to take elastic, and press. Use vertical pins to hold in place. For crisp, easy-to-handle fabric it is possible to machine stitch along the folded edge over the pins. For slippery or heavy fabric it is necessary to tack first before machining. Leave a 1 cm gap in the stitching to put the elastic in.

2 Edge-stitch along the top edge of the fabric to provide a crisp finish and stop the elastic from rolling. Use a safety pin to thread the elastic through the casing. Then adjust the length for comfortable wear. Sew the two ends of the elastic together by oversewing or machining. Stitch along the gap to close up the casing and then across the elastic to anchor it.

Faced edges

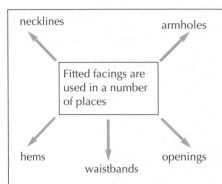

necklines

armholes

Fitted facings are used in a number of places

hems

waistbands

openings

Facings can be cut from:
- the same fabric
- a lining fabric
- a contrasting fabric and applied to the right side as decoration

Vilene is often used to strengthen facings. It is cut out and applied to the wrong side.

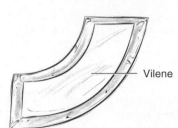

Vilene

Staystitching

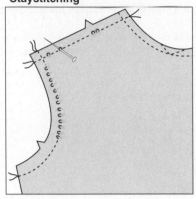

Staystitching prevents stretching, but allows fabric to be eased into a seam line.

Integrated neckline

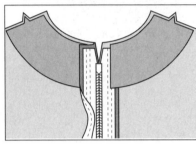

Position cut edges under seam allowances.

Combination facing

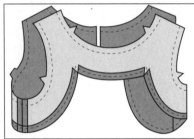

The neck and armholes are finished by the same facing unit.

Separate front facing

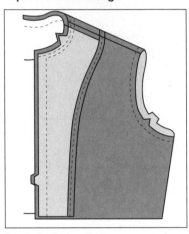

Stitch the facing to the garment along the seam line.

Crossway strips are used as binding and facings in the following ways:
- for neatening seam or hem edges
- for neatening curved raw edges, armholes and neckholes
- as a decorative finish
- for attaching collars, cuffs and frills

Binding an edge

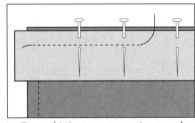

1 Cut and join crossway strips together to provide longer strips of binding as required. Pin to right side of fabric and stitch.

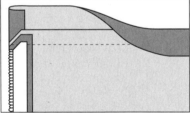

2 When binding is used as a facing edge, first trim the seam allowance off the product, down to 3 mm from the seam line.

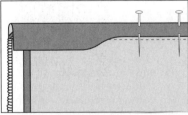

3 Turn up the binding from the stitching line and press in place. Turn to the wrong side of the product and, without stretching it, fold the binding over the raw edge to almost meet the stitching line. Pin tack and press.

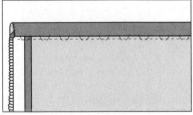

4 Tack, then hem the fold on to the stitching line, remove tacks and press.

Cutting bias strips

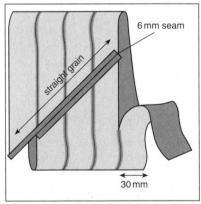

6 mm seam

straight grain

30 mm

Mark 30 mm strips on the fabric, at 45° to the selvedge. Mark 6 mm seam allowance on lengthwise grain. Fold fabric into a tube, right sides together, aligning seams and marked strips. One strip will extend beyond the edge on each side. Stitch; press open. Starting at one edge, cut continuously along the marked line.

Hems and piping

A hem finishes the edge of a garment or product. It is usually worked on the wrong side of the fabric and can also be decorative.

HEMS

Here are three simple hemming techniques. If you are using heavy fabric, or are planning a fairly complex product, you may need to consider other techniques such as adding a binding.

Turned-up hem

Turn up the required hem depth and tack along the fold at the base. Turn under the top edge of the hem and edgestitch if required. Slipstitch the hem edge to the fabric. Remove tacking stitches.

Fusible webbing

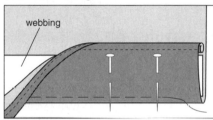

webbing

Iron on the fusible web between the fabric and the hem.

Narrow machined hem

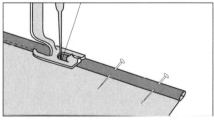

Turn up an 8 mm hem depth and press. Repeat to form a double-turned hem. Pin the hem, matching the seam lines and the grain on the hem and garment to avoid stretching and puckering. Machine stitch close to the inside fold.

Tips for hemming

- Before actually sewing the hem, tack in place, hang up the garment or product, and check that the length and look suit it.
- Allow longer garments and products such as curtains to hang for a few days to let the fabric drop.
- Consider the weight of the fabric – heavier fabrics need deeper hems.
- Choose the method to be used carefully to suit the fabric and purpose of the product.
- Make sure that the stitching is not pulled so tight that it causes puckering.
- Choose a thread that is a shade darker than the fabric – this will blend in well.

Alternative methods for neatening an edge include sewing on a frill (either a bought length or a frill you have made yourself), or creating a piped edge.

A frilled edge

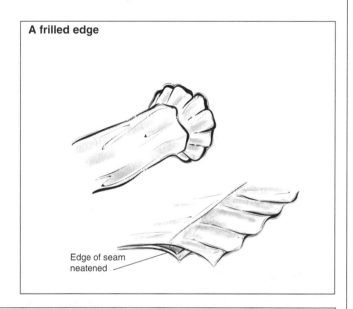

Edge of seam neatened

A PIPED EDGE

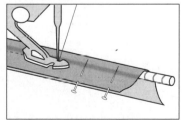

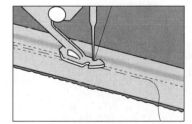

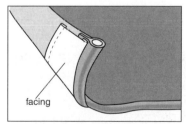

facing

1 Cover the cord with a strip of bias fabric and stitch close to the cord using a cording foot on the sewing machine (or a zipper foot is a cording foot is not available).

2 Tack covered piping to right side of product, and stitch in place.

3 Tack lining or facing to main fabric, with piping sandwiched between the two edges. Stitch along seam line. Turn so that piping lies between edge of fabric and lining or facing. Hemstitch facing with fine stitches to hold it in place.

Sleeves

Sleeves can be an important design feature based around a number of different lengths and styles.

- cap
- short
- elbow
- $\frac{3}{4}$ length
- bracelet
- long

SET-IN SLEEVES

The set-in sleeve is widely used on coats, jackets, blouses, dresses and jumpers.

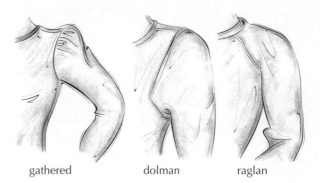

gathered   dolman   raglan

Variations on a set-in sleeve:

- **Gathered sleeve** – fabric fullness evenly gathered over the top of the sleeve with slightly more to the front.
- **Dolman sleeve** – deep armhole and very little ease in the cap of the sleeve. Popular with sweatshirts.
- **Raglan sleeve** – a diagonal seam joins the sleeve and bodice and extends to the neckline.

How to sew a set-in sleeve

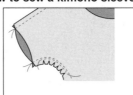

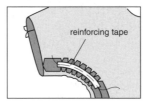

1 Make up the sleeves and finish lower edges. Machine gathering stitches around the sleeve head, between notches.

2 Insert sleeve into armhole with sleeve head to shoulder seam. Match side notches and underarm seams. Pull gathering thread evenly between centre top and side notches. Pin and tuck sleeve in place.

3 Machine stitch around the sleeve on the seam line. Remove gathering and tacking stitches; trim and neaten seam allowances. Press over a sleeve pad, pressing out any fullness around the head.

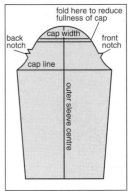

fold here to reduce fullness of cap
cap width
back notch   front notch
cap line
outer sleeve centre

A set-in sleeve pattern

KIMONO AND SHIRT SLEEVES

How to sew a kimono sleeve

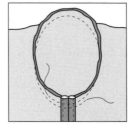

reinforcing tape

The **kimono sleeve** is cut in one piece with the bodice with a generous armhole.

1 Stitch the section of the seam under the arm with a short stitch. Snip into the seam allowance and press the seam open.

2 Cut a 10 cm length of reinforcing tape. Position the

tape centrally over the underarm seam, tuck under the raw ends then pin and tack. Working from the right side, stitch the tape in place by stitching 3 mm on either side of the seam.

Check points for sleeves:
- Sleeves have a front and back – single notch at the front, double notch at the back.
- The sleeve should sit exactly at the top of the shoulder.
- Once inserted into the armhole, the sleeve should hang slightly forward from the shoulder, with the grain of the fabric running from the shoulder to the little finger.

A **shirt sleeve** is stitched at the shoulder with a double stitched seam before the side seam is sewn.

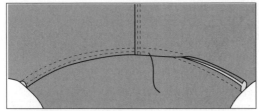

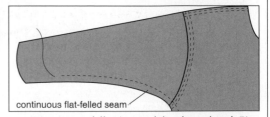

continuous flat-felled seam

How to sew a shirt sleeve

1 Leave the side seam of the garment and underarm seam of the sleeve unstitched. With wrong sides together, pin the sleeve into the armhole, matching the notch to the shoulder seam and

easing in any fullness round the sleeve head. Pin, tack and stitch in place with a flat-felled seam.

2 Refold the shirt and stitch the sleeve seam and side seam of the garment in one continuous stop, using a flat-felled seam.

Collars and cuffs

COLLARS

Choice of collar depends on the shape of the neck edge, the thickness of the fabric and the way the collar will be applied to the neckline of the garment.

The three basic collar types are:

- **Flat** – they lie against the garment and follow the neck edge closely.
- **Stand** – sewn onto a curved neckline with an insert to allow them to stand up from the neckline.
- **Roll** – these have no interfacing and roll over away from the neckline.

Sailor

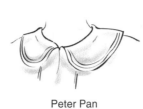
Peter Pan

Tunnel

Tuxedo

Revere

Types of collar

Cowl

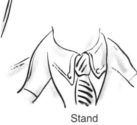
Mandarin

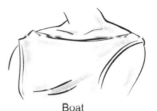
Stand

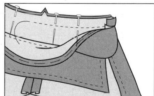
Boat

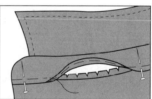
Polo Neck

How to sew a shirt collar

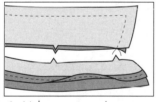 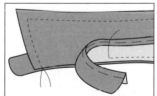  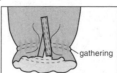

1 Make up a two-piece collar as shown. Topstitch the outer edges of collar as necessary. Turn under one neck edge of collar stand and tack.

2 With right sides together, sandwich the collar between the stand pieces, matching notches with collar ends to centre front. Pin, tack and stitch down the front edges of the stand.

3 Position the unfinished edge of the stand against the wrong side of the garment, matching ends and notches. Pin, tack and stitch it in place. Trim and press the turnings up inside the stand.

4 Tack and stitch the turned-under edge of the stand on the right side, covering the previous stitches.

CUFFS

How to sew a gathered sleeve onto a plain cuff

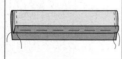 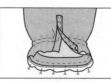 
gathering

1 Cut out a one-piece cuff. Fuse the interfacing to one half of the cuff. Fold the cuff in half, right sides together.

2 Fold up one turning, pin and stitch the ends right up to the seam line. Finish the seam with a few backstitches.

3 Complete the sleeve opening. Work two rows of gathering stitches round the sleeve just inside the seam allowance.

4 Pin the cuff against the right side of the sleeve. Pull up the gathers evenly to fit. Stitch the cuff in place. Trim the turnings and press.

5 Place the remaining edge of the cuff over the previous stitches; pin and tack. Topstitch all around the cuff. Add fastenings.

Pleats and pockets

PLEATS

The 'permanent' pleat arrived with synthetic fabrics which are thermoplastic, that is they can be heat-set into place.

Knife pleats

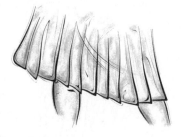

Box pleats

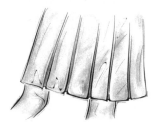

Inverted pleat

Kick pleat

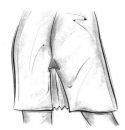

An arrowhead is used to reinforce the seam ending at the top of the pleat.

Stages in pleat production

1 Mark the pleats on the wrong side of the fabric using an appropriate method, e.g. tailor's chalk or fabric marker.
2 For inverted pleats or box pleats, use one colour for the solid lines which mark the fold and a different colour for the dotted lines which indicate where the pleats meet.
3 If using pressed pleats – do the hem next!
4 After marking, place the fabric on a smooth flat surface and pin the pleats into position.
5 Press them carefully in position and then tack them before sewing.

Crinkled fabric

All-over crinkled fabrics are produced by scrunching the fabric into a tube which is then heated.
It is very popular for dresses and skirts.

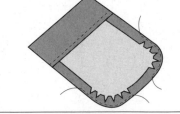

POCKETS

A pocket should be strong and large enough for a purpose. It can be functional and decorative.

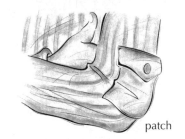
patch

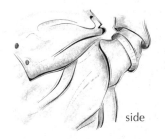
welt

side

HOW TO SEW A PATCH POCKET WITH ROUNDED CORNERS AND SELF-FACING

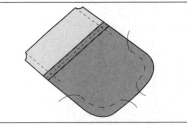

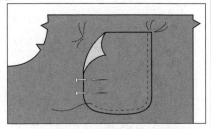

1 Cut out a pocket, turn under the top edge, pin and stitch. Turn the facing back. Pin and stitch the sides; trim and cut diagonally across the top corners. Work a row of stitching around the pocket corners along the seam line.

2 Turn the pocket top right side out, pushing out the top corners. Turn under the remaining edges along the line of stitching. Cut notches into the seam allowance around the curves.

3 Press the pocket flat. Position the pocket on the garment at the marked position. Pin, tack and topstitch it in place.

Fastenings

Fastenings can be functional or decorative. They should:

- be suitable for the purpose
- be the correct size and weight
- be secured firmly
- lie flat

poppers

zips

buckles

Fastenings include

buttons/
buttonholes

hooks and eyes

studs

velcro

BUTTONS

Buttons can be used for decoration as well as for fastenings. When selecting buttons you need to consider:

- Will the finish come off, crack, peel, scratch or discolour with wear?
- Will the button become lighter, darker, cloudy, will it rust or cause fabric abrasion?
- Is the dye colourfast to dry cleaning and/or laundering?
- Will the shank break?

Buttonholes

Modern machines offer a variety of buttonhole styles to suit different purposes and positions. Always practise well before and mark buttonhole positions accurately.

Machined buttonhole

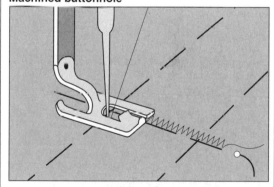

1 Work a close zigzag stitch down the left-hand side of the marked buttonhole. Pivot the fabric and work a few stitches across the end. Pivot again, then work the second side.

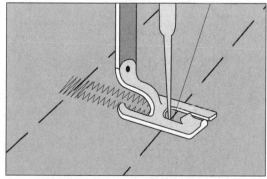

2 At the opposite end, work across the whole buttonhole for a few stitches and fasten off. Carefully cut down the centre.

ZIP FASTENERS

You should choose the correct type and weight for the opening. The types available are:

- lightweight – for dresses, cotton skirts and shorts
- featherweight – for neck and wrist openings
- skirt weight – for skirts, trousers and shorts
- open ended – for jackets, coats and cardigans
- curved – for front openings on trousers and bags
- invisible – for skirts and dresses
- decorative – for dresses, jackets and bags

Inserting a zip

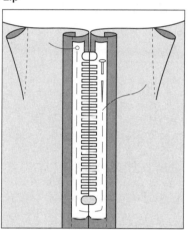

1 Pin and tack the complete seam. Measure and mark the position of the end of the seam. Neaten the edges. Press the entire seam open. Position the zip right side down over the tacked section of the seam, with teeth centrally placed over seam line. Pin and tack in place down each side.

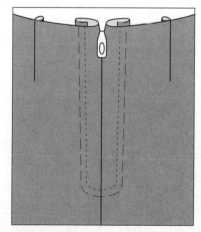

2 Turn over and stitch from the right side following the tacked lines. Use the zipper foot on the sewing machine. Pivot the fabric at the base of the first line of stitching and stitch across the base of the zip. Pivot again and continue stitching up the opposite side. Press and remove tacking stitches.

Using paper patterns

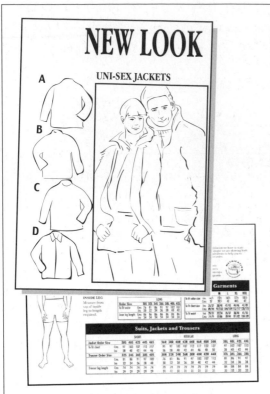

KNOW AND UNDERSTAND PATTERN MARKINGS

Pattern marking or symbol	Meaning
≡≡≡	Lengthen or shorten the pattern here
←——→	The straight of the grain
↓——↓	Lay on the folded edge of the fabric
—✂—	Cutting line
▬▼▼▼▼▼	Position and size of zip opening
– – – –	Fitting or seam line
◆——◆——◆	These show where one piece should join onto another. They are called **balance marks**.
•– – •– – •	Dart
•——————•	Centre of fold line
– – –⌂– – –	Pleat
⊚	Position and size of button
⊢——⊣	Position and size of buttonhole
——————→	Stitch in direction of arrow

Cut out each pattern piece carefully, close to the edge. As you cut place a hand on the pattern piece to keep the fabric flat. Cut with long, sweeping cuts and cut all balance marks outwards.

The front of a paper pattern shows a drawing or photo of the items that can be made. The back shows a chart to calculate fabric requirements, the body measurements for each size and guidelines on suitable fabrics and components to use.

LAYING OUT PATTERN PIECES
Follow the layout guide carefully and make sure that:
- you have got all the pattern pieces required for the garment
- the pattern is the correct size – alter the pattern if necessary
- any large design on the fabric is positioned centre front or back or in the middle of a sleeve
- if the fabric has checks or stripes, these will match at the seams
- if the fabric has a 'nap' – that is, a one-way design or pile – this is allowed for
- pattern pieces with a fold line are placed exactly on the fold
- straight-grain lines on the pattern pieces are parallel to the selvedge edge of the fabric, and lie along the warp threads.

REMEMBER: before removing the pattern pieces, transfer all the necessary pattern marking to the fabric.

Tailor's chalk is one method, tailor tacking is another.

Tailor tacking
1 Stitch through pattern holes leaving loops.

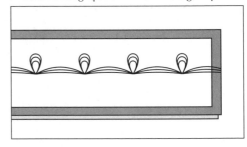

2 Snip between the loops.

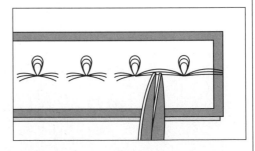

3 Gently pull the layers of fabric apart and cut the threads between them.

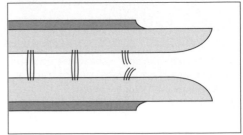

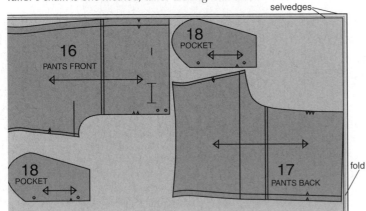

Laying out a pattern

Adapting a pattern

Existing patterns can be made wider or thinner, longer or shorter at any required point.

1 Draw a straight line at the point you wish to alter.

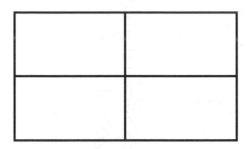

2 To add width or length cut along the line and separate out the pieces to give the required extra centimeters.

3 Place a piece of paper underneath and after checking the measurements pin the pattern pieces to the paper. Use a ruler to draw in new cutting lines.

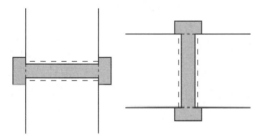

4 To remove width or length, fold the pattern into a pleat along the line. A 2 cm pleat will shorten by 4 cms.

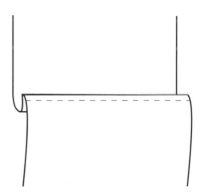

5 Having checked your measurements are correct along the pleat, pin the pleat into position.

To alter a curve

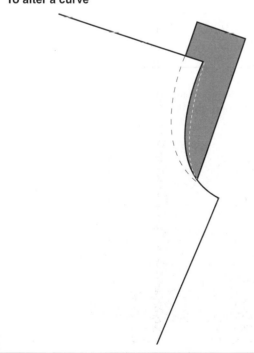

For example – to drop a shoulder line.
1 Place a piece of paper under the pattern and pin the pattern to it.
2 Use a ruler to draw an extended line to give the extra required measurement.
3 Very carefully draw a cutting line to join a suitable point on the curve.

Using dyes

Definitions

Tie-dyeing is a process in which areas of the fabric are tied off according to the design you want to produce.

Batik is a process in which wax is used to create lines and areas resistant to dye.

Tritik is a process that uses stitching gathered up tightly around the design to resist the dye.

Resist dyeing means that the methods used stop the fabric from taking up the colour of the dye.

TIE-DYEING

When tie-dyeing:
- plan your design carefully
- use elastic bands for cold-water dyes
- make sure your design is even over the fabric

Tie-dye designs

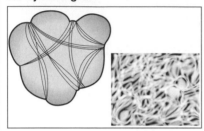

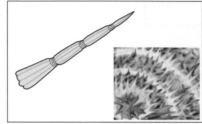

1 Crumple fabric and tie tightly with elastic bands – sometimes known as a marbling effect.

2 Pinch centre of fabric to a point, tie at intervals.

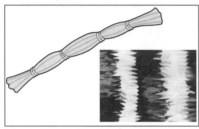

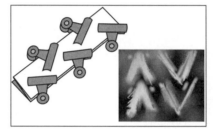

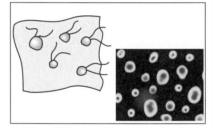

3 Concertina-fold fabric, tie at intervals.

4 Concertina-fold fabric, place pegs or bulldog clips at intervals along folds.

5 Tie pebbles, marbles or buttons into the fabric at intervals.

BATIK

Do take care if you are using hot wax. Achrobatik is an alternative cold wax medium.

Batik process

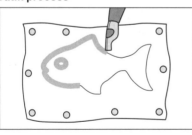

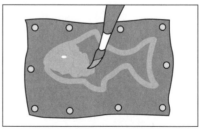

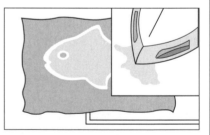

1 Wash and rinse the fabric thoroughly to remove all traces of size. Dry and iron before tacking it to a frame. Draw the design on paper and transfer it to the fabric using a fabric pencil. If the fabric is fine, the design may be visible through the fabric, so no need to mark the fabric at all.

2 Melt the wax and then apply it with a tjanting (tool for applying wax) or a brush. Wax cools more slowly in a tjanting, so use hotter wax for brushwork. Any leftover wax in the brush will re-melt the next time it is used.

3 The fabric is dipped in a dye for each stage of the design; lighter colours must be applied first.
Iron out the wax by placing the dried fabric between sheets of paper towel. Iron with a dry iron set at the next lower setting to the normal one for the fabric. The wax will melt and be absorbed by the towels revealing the colours beneath.

TRITIK

Tritik is generally more effective for large, simple and bold designs. It involves stitching along the outlines of a design drawn on the fabric with a fabric pencil or fabric marker pen. The material is gathered tightly along the stitches and dyed. When the fabric is dry and the stitches are removed the design outline shows up as patches of undyed cloth.

Pens and paints

Fabric pens and paints are now readily available in a full range of colours. They are an easy and quick method of applying colour and design to an otherwise plain fabric.

Before you start, some important points to remember:

- Wash and dry new fabric first to remove any size from it.
- Cover your work surface with a good waterproof cover.
- Protect your own clothes with an apron or old shirt.
- Keep your hands clean, as inky fingers can cause smudges.
- Follow the instructions given by the manufacturer carefully.

Fabric pens and crayons are useful for lines and smaller areas.

Fabric paints can be applied:

- using a large paintbrush for big areas
- with a sponge for a textured appearance
- using an airbrush (see page 44)

STENCILS

Stencils are an easy, quick method of producing a repeat design. Larger stencils can even be cut using a computer-controlled CAMM 1 machine (see page 25).

Fabric paint can be brushed or sponged onto the fabric through the cut-out design.

Always test it to make sure that the paint does not bleed around the edges of the design.

Applying colour using a stencil

Fabric crayons	A very easy method.
Brush	Brushes with soft bristle preserve the stencil better. Use a separate brush for each colour. Toothbrushes and nailbrushes can be used to spot the paint over the stencil by using a flicking action.
Sponge	Soft mottled effects can be created by dabbing the paint on with a sponge.
Roller	Large stencils can be filled in with rollers. Always remove any excess paint from the saturated roller before use.
Airbrushing	This can give a soft, misty appearance.

Special effects

There are a number of special techniques that can be used very creatively for fabric designs.

Using an airbrush

A simpler way is using an airball which is squeezed by hand to spray simple designs.

AIRBRUSHING

When airbrushing:
- Iron the fabric and tape the stencil onto the fabric.
- Have cleaner fluid to hand to clean out the airbrush between colours.
- Hold the airbrush about 10 cm from the fabric and move the airbrush in a steady motion.

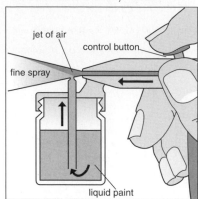

jet of air
control button
fine spray
liquid paint

MARBLING

Marbling requires a base liquid made from water and thickener. The second or third transfer taken is often the best.

Marbling process

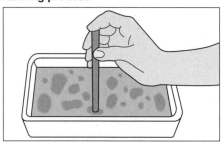

1 Drop paint onto the base liquid using a touch-drop brush.

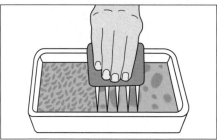

2 Create a marble swirl pattern using a 'comb' (nails in wood).

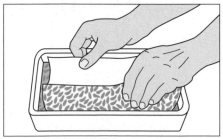

3 Lay the fabric flat on the surface. Lift the fabric from the base liquid very carefully.

SALT DIFFUSION

Salt diffusion works best with fine-textured light-weight fabrics such as silk and fine cottons.

Salt diffusion process

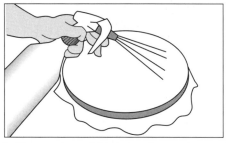

1 Place a piece of fabric in a frame, dampen fabric a little with a spray of clean water (or use a sponge).

2 Use a paintbrush to apply some *diluted* paints (add water). Use a clean brush for each colour.

3 Sprinkle some coarse sea-salt crystals on the wet fabric paint; patterns will form.

Surface decoration

Definitions
Appliqué means applying one fabric on top of another.
Beadwork is the addition of beads, pearls, sequins and rhinestones to fabric.
Image transfer is a modern technique that allows you to transfer images from artwork or photographs onto fabric.
Dimensional fabric paints create a raised finish on fabric.

APPLIQUÉ

Appliqué fabric for your design can be applied using either hand stitching and an appropriate embroidery stitch or machine stitching, plain or embroidery stitch. A quick method is to use Bondaweb.

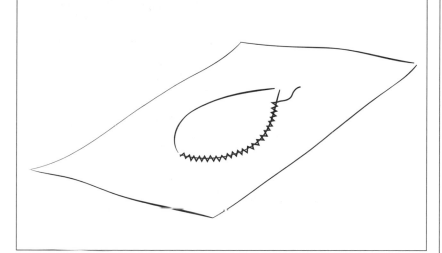

BEADWORK

Beadwork can add inexpensive glamour to any product. There are three main techniques:

Spotting

Outlining

Filling

DIMENSIONAL PAINTS

Dimensional paints are expensive but can add glitter and glamour to fabric.

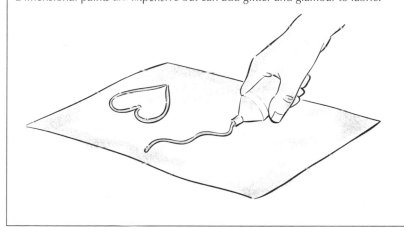

IMAGE TRANSFER

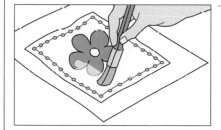 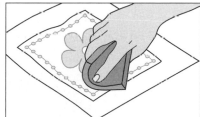 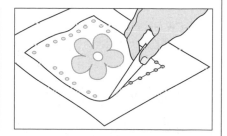

1 Brush the image transfer medium onto the picture or a photocopy of your picture. Place this firmly onto the fabric. Any text must be reversed or removed.

2 When the image transfer medium is dry, soak the picture with a sponge.

3 Pull away the picture to reveal the transferred image. Apply another coat to seal your picture. The result is permanent and machine washable.

Note: Remember that when using certain pictures, **copyright laws** may apply. You may need to ask permission first before using the picture. Always check with your teacher.

Traditional methods of creating decorative fabric

PATCHWORK

Patchwork is a method of producing decorative designs using shaped pieces of fabric. There are many, many traditional patchwork designs.

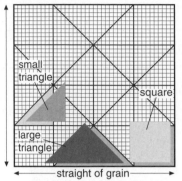

The star design is created using three basic shapes; a square, a large triangle and a small triangle.

The star block is used here as an example of the construction of a simple American patchwork block. Templates are made using squared paper, and these are used to cut out the required number of patches for each block. To make the block, the patches are placed right sides together and the seams sewn with a small running stitch, either by hand or using a machine.

QUILTING

The process of quilting involves sewing together a sandwich of two fabric layers with a centre layer of thick, soft wadding fabric. It is used for insulation and protection against heat (oven glove) and cold (sleeping bag, dressing gown).

This example of machine quilting has the title 'Curl curl beach'.
Free machine quilting is done as follows:

1 Attach a darning foot or 'Big Foot' to the sewing machine; disengaging the feed dogs allows you to move the fabric about.
2 Set the machine to stitch-length 3 and stitch-width 0.
3 Place the fabric in the hoop under the foot.
4 Stitch freely by swinging the hoop about to make a pattern of lines that gradually builds outwards.
5 Move the hoop across the fabric and continue your line.

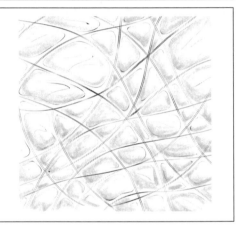

EMBROIDERY

It is best to use stranded cotton and a sharp, crewel needle.

Most sewing machines now also provide a range of machine embroidery stitches.

A range of hand embroidery stitches are shown here.

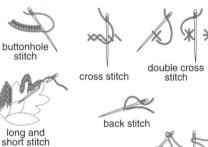

lazy daisy stitch

seed stitch

double satin stitch

chain stitch

satin stitch

buttonhole stitch

cross stitch

double cross stitch

long and short stitch

back stitch

stem stitch

bullion or "grub" rose

bullion stitch

herringbone stitch

french knot

BLOCK PRINTING

Block printing is a process in which paint or dye is applied to a pattern shape on the block which is then pressed onto the fabric. It is still used in many countries to create traditional patterns.

1 Cut a shape out of card or plastic. Glue the shape onto thick card or thin wood. Details can be scored into the plastic.

2 Dip block into a tray of paint, or use a printing roller to cover with paint, then press the printing block firmly on the fabric.

Scales of production

There are three main types of textile production.

	Individual production Job production Make-through production	Batch production	Mass production
Quantity	Each product is made once or in small quantities	Products made in specific quantities. May be one production run or repeat runs at certain times. Can range from 2/3 products per run to 100 000	Continuous vast quantities of products produced
Skills/machinery	Highly skilled, flexible operators and versatile machinery	Flexible operators and versatile machines to meet a wide range of products and processes	Machinery and operator skills have to be highly specific for the job
Product types	Tailored suits Wedding dresses Specialized fabric printing Craft products Prototypes/sample products	Fabric print runs Garment production to meet seasonal changes, for example, Christmas products	Dyeing fabric Hosiery

Main production stages

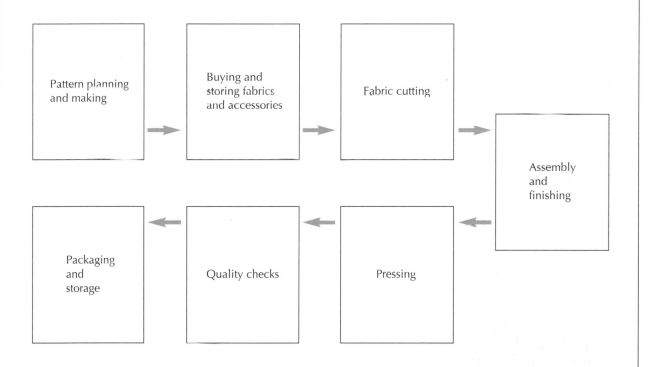

Other important production systems include the following:

• **Synchronized systems** are straight-line work which flows with timed, repetitive operations.

• **Progressive bundle systems** use teams to produce parts of a garment or product, which is then passed on to the next team.

• **Section systems** or **quality circles** use small teams of highly skilled, flexible operators to produce complete products. (Also known as cell production teams.)

Developing pattern design

The modern design room has turned to computer-aided design systems to provide rapid responses and improved styling techniques.

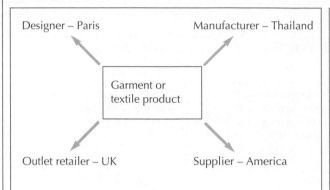

Designer – Paris   Manufacturer – Thailand

Garment or textile product

Outlet retailer – UK   Supplier – America

Manufacturing organizations can now be worldwide: all of it is computer controlled.

INTELLIGENT DRAWING SYSTEMS

An ergonomically designed work station and software allows the designer to quickly develop pattern piece designs to meet product requirements.

INTEGRATED DATABASE

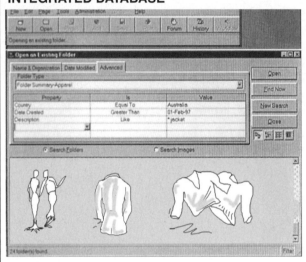

This contains all the information on costings, garment design, trims and components, etc. Digitized images aid selection.

SIMULATION

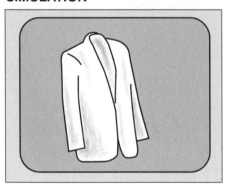

An 'image of reality' is a 3D wire model of the body form enhanced with digitized garment and fabric designs to simulate the garments being worn. Also now available with a catwalk format.

PATTERN GRADING

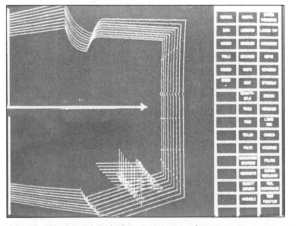

A basic pattern is graded into a series of sizes using computer software programmed to follow defined rules on statistics and measurements.

PATTERN LAY OR CUTTING PLANS

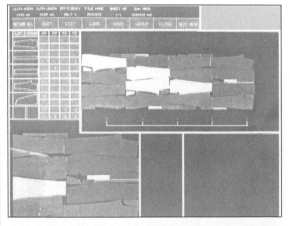

Operators use touch-screen pens to pick up, move, rotate and place pattern pieces on the fabric areas of the lay plan simulation. The software keeps them updated on potential wastage percentages as they position each pattern piece. The lay cutting plan is calculated to make the greatest use of fabric area and to keep waste fabric to less than 5%.

Materials handling in industry

Modern systems reduce materials handling as much as possible as it does not actually add any value to the final product. Machinery is used throughout production.

Robotic carousel loading system for moving rolls of fabric from the storage area to the spreading machine.

Computerized spreading machines lay the fabric as required in plies (or layers) ready for cutting.

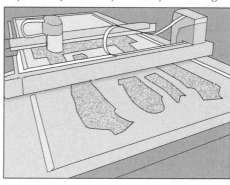

Cutting machines cut the fabric plies into pieces. The cutting is computer controlled and follows the designed pattern lay for the product. A paper pattern is not necessary.

Position marks are made on all layers of the cut fabric pieces using thread markers, drills or hot notchers.

The fabric pieces are separated into bundles to be sent to locations for sewing. This is difficult for automation and so tends to be labour intensive.

Assembly machining processes are designed to machine and assemble specific aspects of the product.

More and more manufacturers are turning to flexible machine assembly systems which can be used for a variety of sewing applications, quickly converted to suit the changing needs of the product batch scheduled each day.

The modern assembly process

Production methods must be flexible in order to respond quickly to changing demand. This means:

- being able to handle different production processes at the same time
- parallel channelling of different batches of product styles
- flexibility of skills

The Eton 2002 system uses real-time software to:

- calculate labour costs
- collect work-in-process information
- balance the production line
- sort work by colour, size, production order and customer priority
- present the product to each operator offering the best working position
- shorten or lengthen the stopping position where different styles and sizes are required
- automatically move work around the shop floor

Machine functions	Applications	Features
Lockstitch	Seams	Stitch the same on both sides
Chain stitch	Seams	Suitable for stretchy fabrics, tacking and basting
Blind stitch	Blind stitch work and hemming	Stitch cannot be seen on the right side of the fabric
Linking	Attaching trimmings and cuffs to knitted fabrics	
Overedge (overlocker)	Edge neatening	Very clean seam finish – built-in trimming device
Flat seam	Binding cut edges/flat seams on knitted fabrics	Two-and three-thread systems
Specialized	Button, buttonhole, bartack	Automated control and specialist attachments

A small hand-held operator terminal is positioned at each work station on the line. At the end of each process, the operator punches in a code, recording the work done and any problems as they arise, e.g. changing the bobbin on the machine. This allows the computer-controlled system to monitor all work done.

Pressing in industry

Pressing techniques involve the application of heat and pressure for short periods of time. Steam, compressed air and suction are all used in various pressing methods.

A steam dolly is used to press skirts, trousers and dresses. The dolly is inflated inside the clothes before pressing.

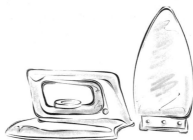

Electric iron

A sleeve press

Moveable flat bed press

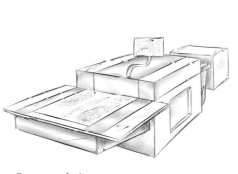

Conveyor fusing press

Pressing	Applications
Under pressing	For example, pressing seams before hemming.
Top pressing	The final finishing pressing techniques.
Flat pressing	Smooths fabric before laying, cutting or sewing, for example to press the seams.
Steaming	• relaxes the fabric • avoids fabric shrinkage
Moulding	Used to create a shape in a product by: • stretching over rounded shapes, for example at the shoulder, bust or corner • pressing to reduce extra gathered fabric, for example at the head of a sleeve or at a corner
Steam dolly	Specialist machine used for skirts, trousers and dresses. 1 The garment is placed over a form and fixed into position at set points. 2 It is inflated for several minutes, using steam and air. 3 While in this position, the seams may be ironed by the operator using a steam iron. 4 The garment is released and removed.
Tunnel finisher	Specialist machine used for shirts and blouses. 1 The garments are positioned on hangers. 2 They pass into the chamber or tunnel, in which they are steamed and dried. 3 Some small parts, such as collars, may need pressing with a steam iron for a better finish.

Final inspection and warehousing

The final quality check on the product is performed by an operator who looks for:

- Fabric – accidental snips or pulls in processing
- Seams – puckering, openings, pulls, finish
- Hems – straight, flat, uniform depth, finish
- Fastenings – zips open and close, sit correctly, buttons pass through buttonholes, etc.
- Pressing – all parts correctly pressed
- Trailing threads – removed using a small pair of hand scissors (snippers)

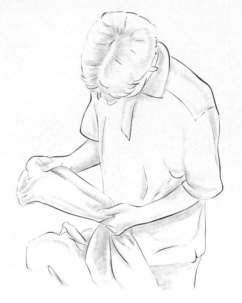

Snippers

Modern warehouse systems ensure that textile items are handled, stored and transported in a way that reduces creasing, so that they arrive at the retailers ready for display.

Just-in-time scheduling means that goods are produced ready to go out to the retailer almost as soon as they have been completed.

Garments are now stored and transported on hangers on moveable rails.
Other products are usually boxed for transport.

Packaging and security

Packaging performs a number of functions.

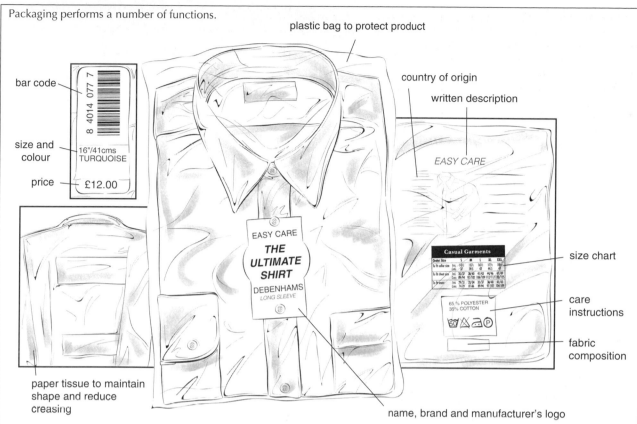

- plastic bag to protect product
- bar code
- size and colour
- price
- country of origin
- written description
- size chart
- care instructions
- fabric composition
- paper tissue to maintain shape and reduce creasing
- name, brand and manufacturer's logo

16"/41cms
TURQUOISE

£12.00

EASY CARE

THE ULTIMATE SHIRT

DEBENHAMS
LONG SLEEVE

EASY CARE

Casual Garments

65% POLYESTER
35% COTTON

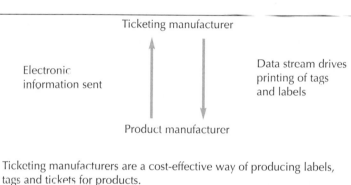

Ticketing manufacturer

Electronic information sent

Data stream drives printing of tags and labels

Product manufacturer

Ticketing manufacturers are a cost-effective way of producing labels, tags and tickets for products.

78019823 124578131

Bar code systems are used to automatically count product sales at the till, re-order stock, clear the product at the warehouse and transport new stock to the retailer.

News Weekly

Parents angry as local comprehensive patents new school bags and limits outlets for school uniforms

Angry parents last week called a meeting with local councillors and the school management of Buckley Park High to protest about new arrangements for bags and uniforms. They allege that the school is aiming to profit unreasonably from sales of bags and clothing from a very limited number of shops.

Intellectual property rights are now used by designers to protect exclusive designs against unlicensed copiers.

Product cycles

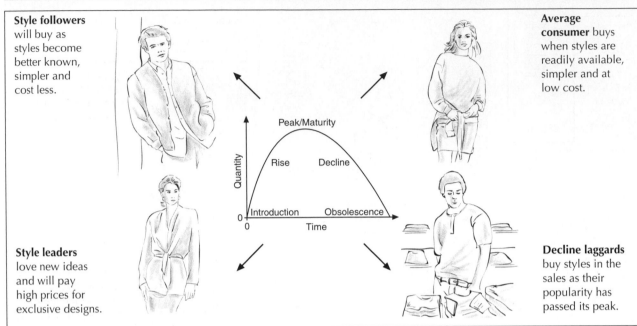

Style followers will buy as styles become better known, simpler and cost less.

Average consumer buys when styles are readily available, simpler and at low cost.

Style leaders love new ideas and will pay high prices for exclusive designs.

Decline laggards buy styles in the sales as their popularity has passed its peak.

DIFFERENT PRODUCT CYCLES

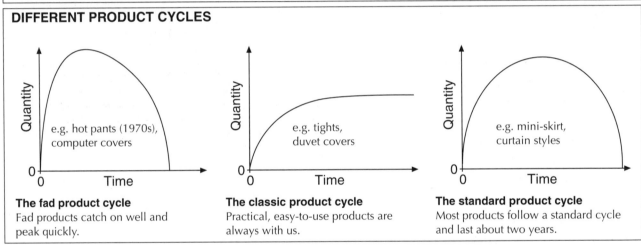

e.g. hot pants (1970s), computer covers

e.g. tights, duvet covers

e.g. mini-skirt, curtain styles

The fad product cycle
Fad products catch on well and peak quickly.

The classic product cycle
Practical, easy-to-use products are always with us.

The standard product cycle
Most products follow a standard cycle and last about two years.

INFLUENCES ON FASHION

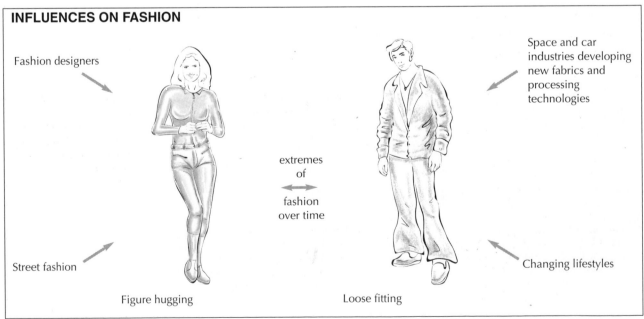

Fashion designers

Space and car industries developing new fabrics and processing technologies

extremes
of
fashion
over time

Street fashion

Changing lifestyles

Figure hugging

Loose fitting

Fashion producers

Charles Frederick Worth changed the face of fashion when he produced the first ever collection of sample garments to show to his customers using models.

Modern fashion designers take inspiration for their designs from many sources:

Other periods of history – also known as revivalist fashion, e.g. 1920s flapper style

Modern media – TV, films, magazines. An example is the growing fashion of kilts for men influenced by the film *Braveheart*.

Influences on fashion designers

New technologies and fabrics open up new ideas, e.g. Gore-Tex

Street fashion – developed by teenagers initially but picked up in the fashion world, e.g. punk

FASHION DESIGNERS OF TODAY

Couturier/ière	Fashion house	City
John Galliano	Christian Dior	Paris
Alexander McQueen	Givenchy	
Christian Lacroix	Christian Lacroix	
Karl Lagerfield	Chanel	
Jean Paul Gaultier	Jean Paul Gaultier	
Yves Saint-Laurent	Yves Saint-Laurent	
Richard Tyler	Richard Tyler Couture	New York
Calvin Klein	Calvin Klein	
Vivienne Westwood	Vivienne Westwood	London
Bruce Oldfield	Bruce Oldfield	
Jasper Conran	Jasper Conran	
Donatella Versace	Versace	Milan
Giorgio Armani	Armani	
Issey Miyake	Issey Miyake	Tokyo

Ready to wear (prêt-a-porter) – not exclusive but affordable, well designed clothes for customers who can afford to pay for them

Bespoke – made for individual clients, usually very expensive

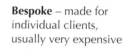

Fashion product ranges

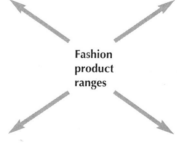

Industrial manufacture – mass-market ranges at affordable prices

Diffusion – range of designer clothes available in the high-street fashion boutiques and department stores at affordable prices

Prices will depend on:
- exclusivity
- design details included
- fabrics/components used
- batch number produced
- percentage of hand work required, etc.

Textiles and the car industry

The use of textiles in car manufacture has grown rapidly for one simple reason – weight! Textiles are far lighter than plastics, which means a greater distance travelled per litre of petrol.

Insulation to reduce vibration and noise
- boot
- floor
- engine
- body panels

Mouldable body parts
- pre-formed interior trim components

Aesthetic features
- carpets
- seat covers
- door trims

Filtration
- to filter air coming into the car
- to filter exhaust fumes from the car to meet environmental standards on pollution

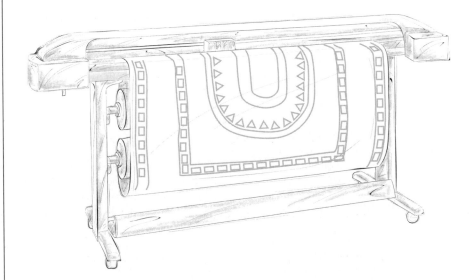

Fabrics used
- needle bonded and thermal bonded non-woven fabrics for insulation
- polyester fabric for car seats and door trims
- Kevlar is used for brake linings

The ENCAD 1500 TXtm printer is a small jet printer specifically designed to print textile samples in a range of colours and designs.

Inkjet printing will allow cars to be personalized at a minimum cost – choose your patterns for car seat fabric, etc.

Presentation of a project

First impressions for the written folio content of your coursework are all-important as a well presented project makes a positive impact.

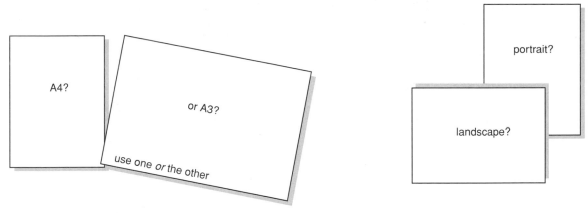

Begin by deciding on the size and orientation of the paper for your folio.

Bindings

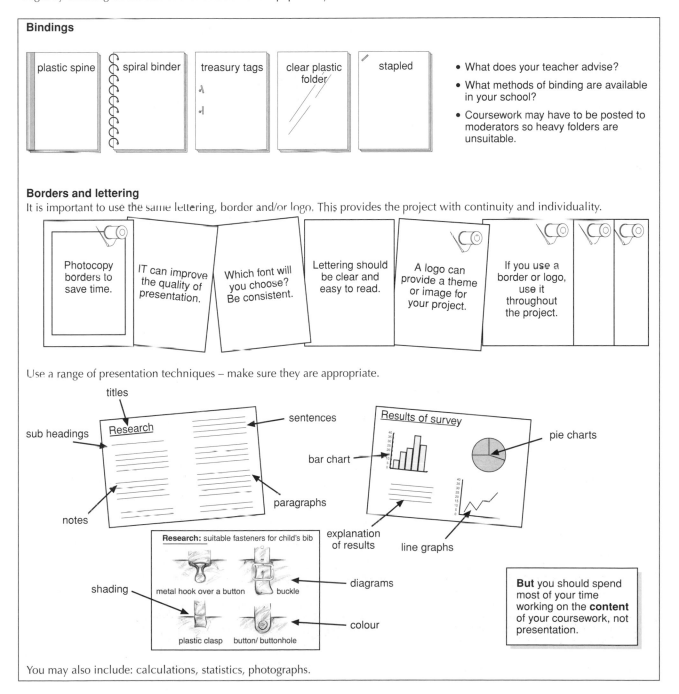

plastic spine | spiral binder | treasury tags | clear plastic folder | stapled

- What does your teacher advise?
- What methods of binding are available in your school?
- Coursework may have to be posted to moderators so heavy folders are unsuitable.

Borders and lettering

It is important to use the same lettering, border and/or logo. This provides the project with continuity and individuality.

Photocopy borders to save time.

IT can improve the quality of presentation.

Which font will you choose? Be consistent.

Lettering should be clear and easy to read.

A logo can provide a theme or image for your project.

If you use a border or logo, use it throughout the project.

Use a range of presentation techniques – make sure they are appropriate.

titles
sentences
sub headings
Research
notes
paragraphs

Results of survey
bar chart
pie charts
explanation of results
line graphs

Research: suitable fasteners for child's bib
metal hook over a button
buckle
shading
plastic clasp
button/ buttonhole
diagrams
colour

But you should spend most of your time working on the **content** of your coursework, not presentation.

You may also include: calculations, statistics, photographs.

Choosing and organizing a project

What have you been asked to do? Have you been:
- given a coursework brief?
- asked to choose from a selection of ideas?
- asked to devise your own context and brief?

REMEMBER
What you choose has to:
- allow you to be able to show your best work
- be of interest to you
- allow you to complete it within the time set
- show what you have learnt and understood in lessons

What do I do now?

People – who are the best people for you to design for?
- children
- teenagers
- adults
- elderly
- special interest or hobby group
- Consider who you have easy access to for trialling, etc.

Cost and resources – how much will you need for fabrics, components, etc., who will pay, what is realistic for your situation? If money and resources are limited – keep it SMALL!

How good are you/how quick are you?
The more complex the product, the more you have to squeeze into the time period in the way of research, processes, finishing, etc. If in doubt keep it SIMPLE AND BOLD!

Evaluation – again, who can help you evaluate your project and the final product?
- local shopkeepers
- manufacturers
- services, e.g. health visitor
- Find a named person with whom you can discuss your project at set points from start to finish.

ORGANIZING YOUR PROJECT FOLIO
Plan a logical sequence for your design folio in advance.

1	Attractive front cover	2	Need identified	3	Research available products	4	Research questionnaire survey	5	Research information gathered
6	Analysis of results Conclusions	7	General specification	8	Initial design ideas	9	Developing those ideas	10	Results of testing and evaluation
11	Developing aspects, e.g. fastenings	12	Testing /evaluating those aspects	13	Product specification	14	Planning Production	15	Testing/ evaluating processes
16	Identifying critical control points	17	Developing a quality assurance system	18	Tests for quality	19	How the product can be produced industrially	20	Testing with customer group
21	Developing ideas further	22	Final annotated solution and packaging ideas	23	Testing final idea	24	Consumer evaluation	25	Final evaluation

Showing evidence of industrial practice

CADCAM – show clearly at least **one** area where you are using CADCAM to produce an aspect of your design a number of times, consistently to a high quality. For example:
- Embroidered logo or design
- A stencil for screenprint work on fabric and packaging materials using a CAMM 1 or other cutter plotter machine
- Packaging materials

Production processes – demonstrate consistent quality in all processes used and their suitability for larger-scale production. For example:
- Seam production
- Fabric printing processes
- Use of fastening processes such as machined buttonholes

Timing – identify stages of production where production timing is controlled. Explain how the process is controlled and why time control is important to this process. For example:
- Drying times for printing processes
- How embroidery design can reduce CADCAM production times by the use of few colours, simpler shapes, etc.
- Production times for insertion of types of fastening can be reduced by suitable choices

Other IT use, for example:
- Use of graphic software for fabric design and stencils, product design ideas, pattern piece development and packaging
- Databases on fabric properties and costs
- Spreadsheets for evaluation of products and analysis of consumer survey results in graphical format
- Desk top publishing (DTP) software to produce well presented design folios

Market research and evaluation – show evidence of the market research you have carried out and the results of any trialling of available products and your final product with an identified customer group.

An identified **quality assurance system** for your product. Consider:
- The main areas where quality could fail
- What needs to be checked to maintain quality
- How these identified features can be tested and the criteria for quality

Quality control checks – show evidence of:
- The testing procedures you have used
- The results of tests you have carried out
- How well they match your identified criteria for quality

Packaging – show evidence that you understand the role of packaging for manufactured textile products by discussing packaging for your product in terms of:
- Suitability of materials
- Transportation
- Display
- Information for the customer
- Barcoding systems
- Anti-theft system

N.B. It is not necessary to actually produce packaging for the product.

Showing evidence of systems and control

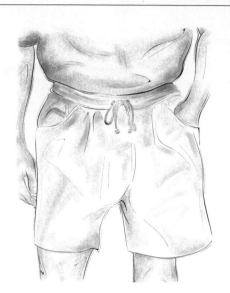

It is important to show clear evidence of your production system and control procedures within your project folio.

Things can go wrong at various points throughout production. Make sure that you clearly identify the points where you will carry out checks and tests, and feedback loops back to a point where you can look at alternative choices.

Highlight points in the production system where you use CADCAM. Explain clearly why the processes you have chosen are suitable for larger scale production.

A PRODUCTION SYSTEM FOR SUMMER SHORTS

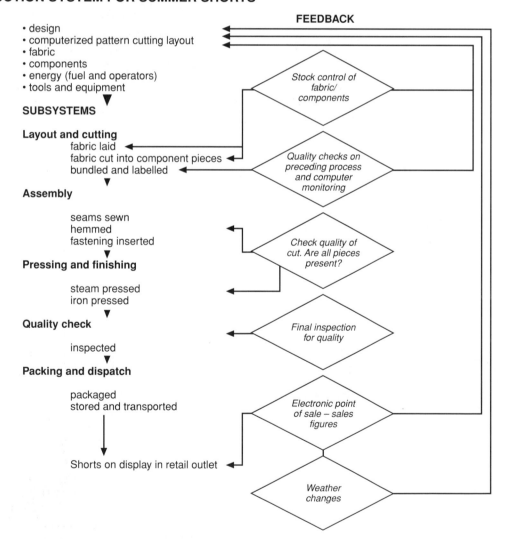

FEEDBACK

- design
- computerized pattern cutting layout
- fabric
- components
- energy (fuel and operators)
- tools and equipment

SUBSYSTEMS

Layout and cutting
　　fabric laid
　　fabric cut into component pieces
　　bundled and labelled

Assembly

　　seams sewn
　　hemmed
　　fastening inserted

Pressing and finishing

　　steam pressed
　　iron pressed

Quality check

　　inspected

Packing and dispatch

　　packaged
　　stored and transported

Shorts on display in retail outlet

Stock control of fabric/components

Quality checks on preceding process and computer monitoring

Check quality of cut. Are all pieces present?

Final inspection for quality

Electronic point of sale – sales figures

Weather changes

Sources of inspiration

The world around us provides many opportunities for inspiration for ideas for design.

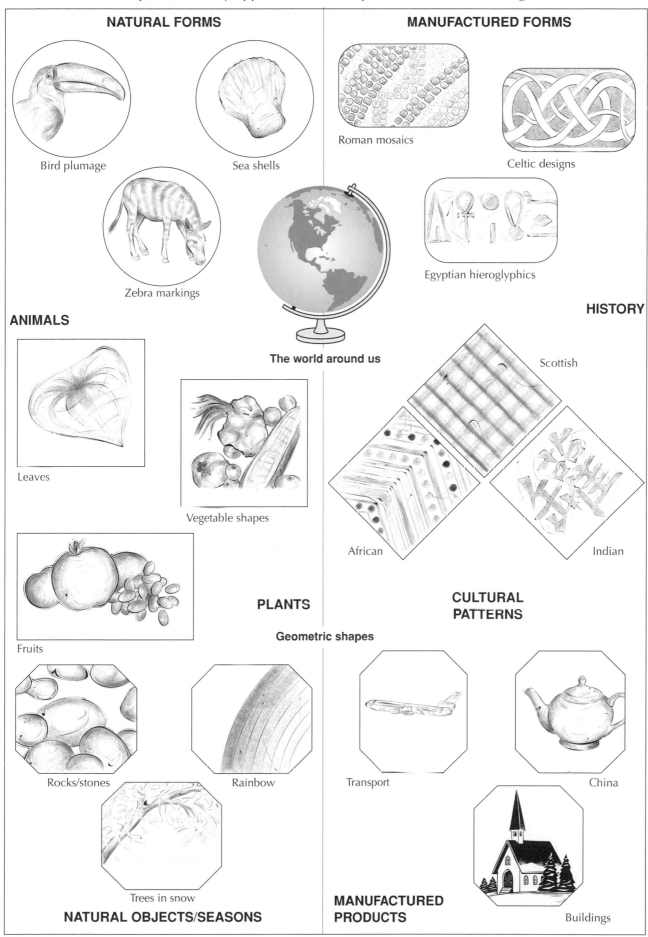

NATURAL FORMS

Bird plumage

Sea shells

Zebra markings

ANIMALS

Leaves

Vegetable shapes

Fruits

Rocks/stones

Rainbow

Trees in snow

PLANTS

Geometric shapes

NATURAL OBJECTS/SEASONS

The world around us

MANUFACTURED FORMS

Roman mosaics

Celtic designs

Egyptian hieroglyphics

HISTORY

Scottish

African

Indian

CULTURAL PATTERNS

Transport

China

Buildings

MANUFACTURED PRODUCTS

Gathering research information

Before you begin designing you must clearly identify what information you need at your fingertips!

KEY WORDS

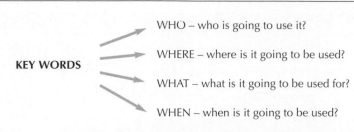

WHO – who is going to use it?

WHERE – where is it going to be used?

WHAT – what is it going to be used for?

WHEN – when is it going to be used?

CUSTOMER GROUP

Talk
Question
Observe

What do they need and why?

EXISTING PRODUCTS

Study
Analyse
Compare
Evaluate in use

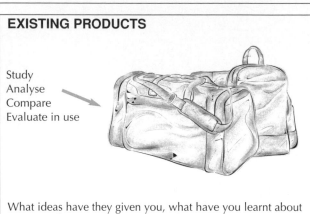

What ideas have they given you, what have you learnt about their use?

EXPERIMENTAL WORK

Materials ...

... and processes

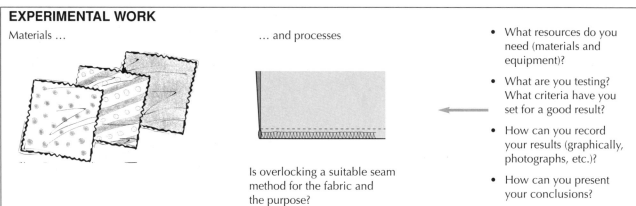

Is overlocking a suitable seam method for the fabric and the purpose?

- What resources do you need (materials and equipment)?
- What are you testing? What criteria have you set for a good result?
- How can you record your results (graphically, photographs, etc.)?
- How can you present your conclusions?

INFORMATION

People/manufacturers/charities
Internet/CD-ROMs
Books/magazines/leaflets

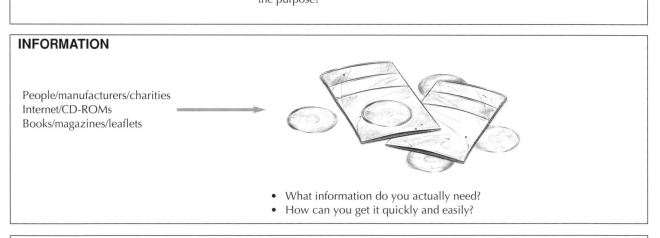

- What information do you actually need?
- How can you get it quickly and easily?

READ	– read all the information carefully and extract what is useful.
RELEVANT	– make sure that all your research work is relevant; good planning should ensure that.
RESOURCES	– identify what resources you need to carry out your research work and if need be develop questionnaires and surveys beforehand.
REPORT	– report the facts clearly and precisely.
REPRESENT	– make sure that your conclusions are represented carefully and clearly in line with the criteria you are looking for.

Analysing and using research material logically

As you carry out your research you need to analyse it carefully as you complete each section. You will then have clear information at your fingertips for your design ideas.

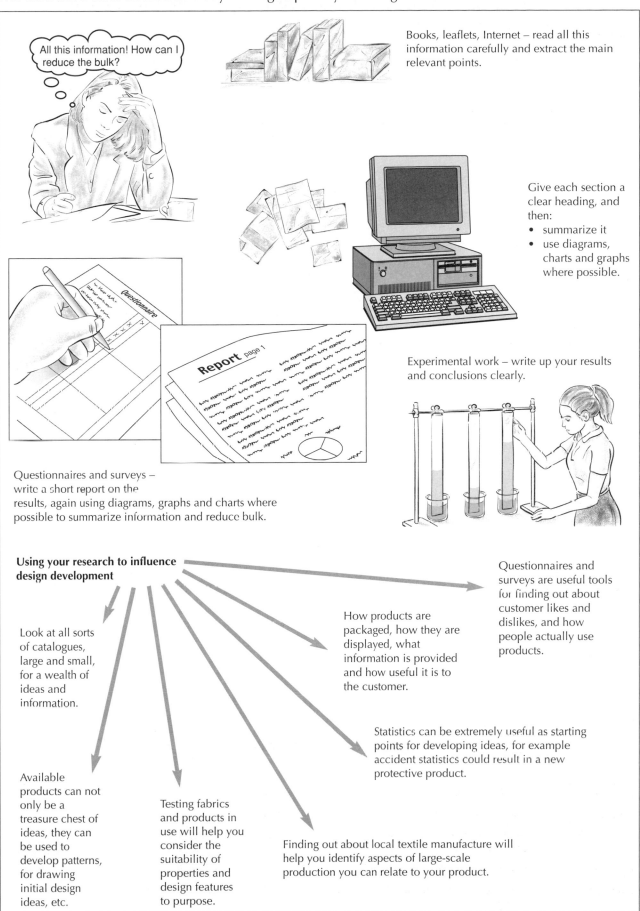

All this information! How can I reduce the bulk?

Books, leaflets, Internet – read all this information carefully and extract the main relevant points.

Give each section a clear heading, and then:
- summarize it
- use diagrams, charts and graphs where possible.

Experimental work – write up your results and conclusions clearly.

Questionnaires and surveys – write a short report on the results, again using diagrams, graphs and charts where possible to summarize information and reduce bulk.

Using your research to influence design development

Look at all sorts of catalogues, large and small, for a wealth of ideas and information.

Available products can not only be a treasure chest of ideas, they can be used to develop patterns, for drawing initial design ideas, etc.

Testing fabrics and products in use will help you consider the suitability of properties and design features to purpose.

How products are packaged, how they are displayed, what information is provided and how useful it is to the customer.

Questionnaires and surveys are useful tools for finding out about customer likes and dislikes, and how people actually use products.

Statistics can be extremely useful as starting points for developing ideas, for example accident statistics could result in a new protective product.

Finding out about local textile manufacture will help you identify aspects of large-scale production you can relate to your product.

Analysing a garment

If you have chosen to design and make a fashion garment or a theatrical costume, it is essential to analyse a full range of available products of a similar type. Look for how other designers have built features into the design, and consider how you can develop these ideas in your design.

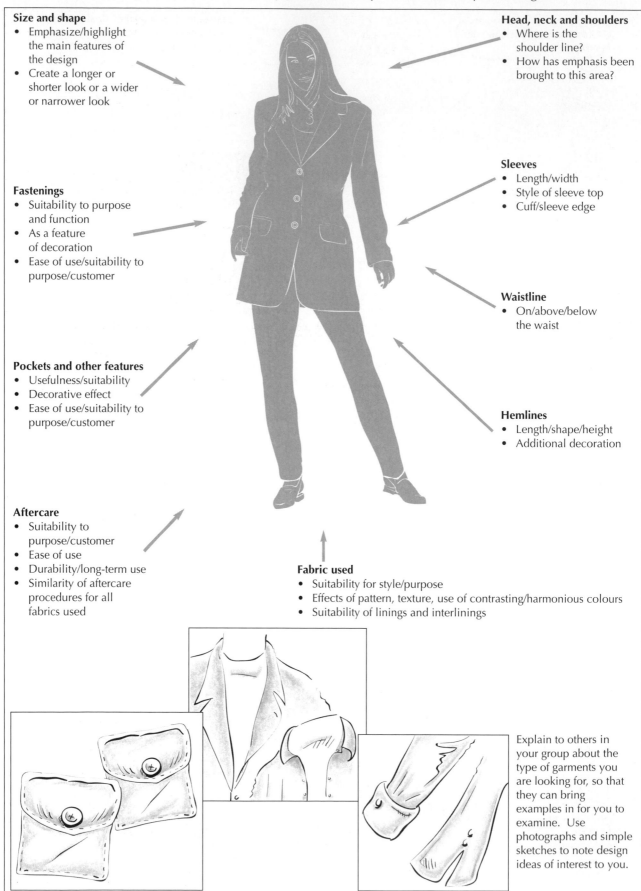

Size and shape
- Emphasize/highlight the main features of the design
- Create a longer or shorter look or a wider or narrower look

Fastenings
- Suitability to purpose and function
- As a feature of decoration
- Ease of use/suitability to purpose/customer

Pockets and other features
- Usefulness/suitability
- Decorative effect
- Ease of use/suitability to purpose/customer

Aftercare
- Suitability to purpose/customer
- Ease of use
- Durability/long-term use
- Similarity of aftercare procedures for all fabrics used

Head, neck and shoulders
- Where is the shoulder line?
- How has emphasis been brought to this area?

Sleeves
- Length/width
- Style of sleeve top
- Cuff/sleeve edge

Waistline
- On/above/below the waist

Hemlines
- Length/shape/height
- Additional decoration

Fabric used
- Suitability for style/purpose
- Effects of pattern, texture, use of contrasting/harmonious colours
- Suitability of linings and interlinings

Explain to others in your group about the type of garments you are looking for, so that they can bring examples in for you to examine. Use photographs and simple sketches to note design ideas of interest to you.

Analysing a textile product

If you have chosen to design and make a textile product, again it is essential to analyse a full range of available products of a similar type. Look for how other designers have built features into the design and consider how you can develop these ideas in your design.

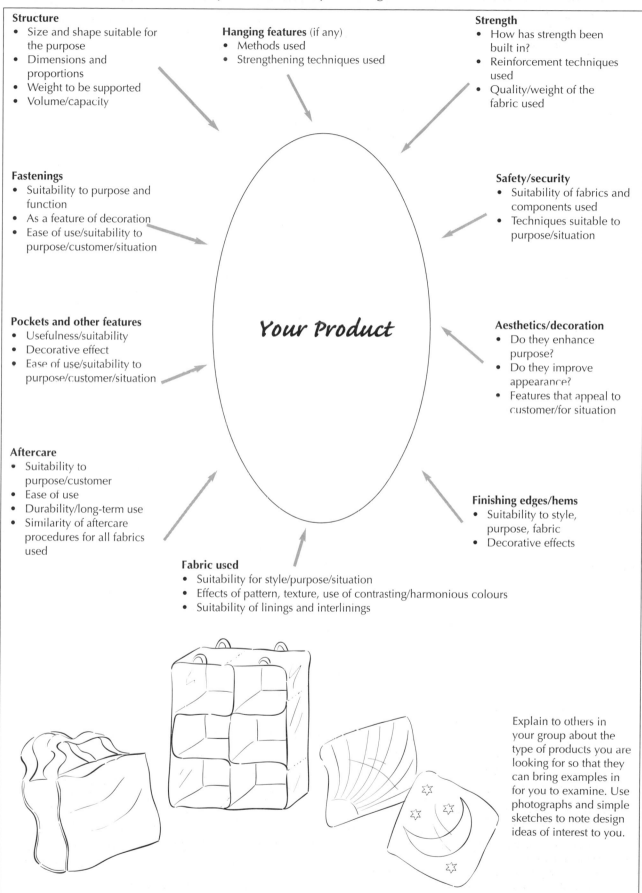

Structure
- Size and shape suitable for the purpose
- Dimensions and proportions
- Weight to be supported
- Volume/capacity

Hanging features (if any)
- Methods used
- Strengthening techniques used

Strength
- How has strength been built in?
- Reinforcement techniques used
- Quality/weight of the fabric used

Fastenings
- Suitability to purpose and function
- As a feature of decoration
- Ease of use/suitability to purpose/customer/situation

Safety/security
- Suitability of fabrics and components used
- Techniques suitable to purpose/situation

Pockets and other features
- Usefulness/suitability
- Decorative effect
- Ease of use/suitability to purpose/customer/situation

Aesthetics/decoration
- Do they enhance purpose?
- Do they improve appearance?
- Features that appeal to customer/for situation

Your Product

Aftercare
- Suitability to purpose/customer
- Ease of use
- Durability/long-term use
- Similarity of aftercare procedures for all fabrics used

Finishing edges/hems
- Suitability to style, purpose, fabric
- Decorative effects

Fabric used
- Suitability for style/purpose/situation
- Effects of pattern, texture, use of contrasting/harmonious colours
- Suitability of linings and interlinings

Explain to others in your group about the type of products you are looking for so that they can bring examples in for you to examine. Use photographs and simple sketches to note design ideas of interest to you.

Developing a brief

TACKLING A BRIEF

WHO?

Who is the product for?

e.g. children   adults   elderly people

WHERE?

Where will it be used?

e.g . classroom   home   garden

WHAT FOR?

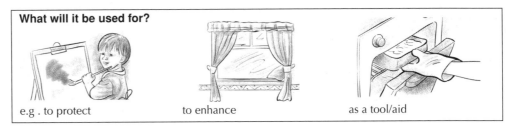

What will it be used for?

e.g . to protect   to enhance   as a tool/aid

WHEN

When will it be used?

e.g . seasonal   special occasion   permanent

Context
Once you can answer these three questions you will have a clear **context** for your brief.

SPECIFICATIONS

These are the details to guide your thinking for both your **design** and **making** ideas. It is important to clearly identify measurements and features you can test for.

For example:
- Must weigh no more than 2 kg.
- Must be no longer than 30 cm.
- Must be able to carry equipment weighing 1.5 kg.
- Must be able to extend to a height of 2 m.
- The size of the stencilled pattern on the fabric must fit comfortably across the width and height of the sides of the box.
- The fastening must be suitable for use by someone with arthritic hands.
- The fabric must not be hazardous for use with young children.

TASK

To design and make a product to meet the context and both the design and manufacture specifications.

- Research and gather information from a variety of sources.
- Develop design ideas and then a final design drawing with annotation showing measurements and construction details.
- Research suitable methods for production.
- Make a suitable prototype product to meet the specifications.
- Continually test throughout against the specifications and finally evaluate your finished product within a working context.

Writing a specification

PERFORMANCE SPECIFICATION

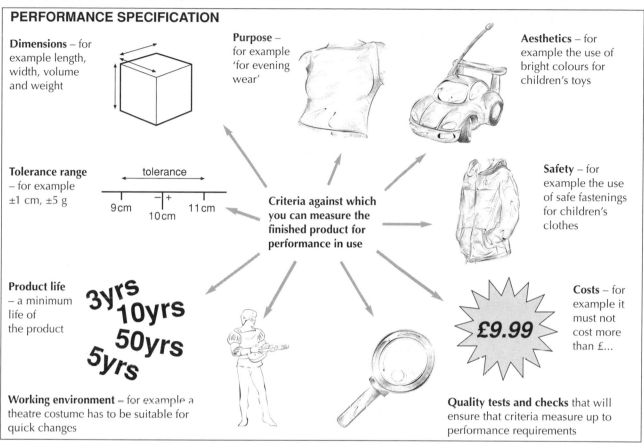

Dimensions – for example length, width, volume and weight

Purpose – for example 'for evening wear'

Aesthetics – for example the use of bright colours for children's toys

Tolerance range – for example ±1 cm, ±5 g

tolerance

9cm 10cm 11cm

Criteria against which you can measure the finished product for performance in use

Safety – for example the use of safe fastenings for children's clothes

Product life – a minimum life of the product

3yrs 10yrs 50yrs 5yrs

£9.99

Costs – for example it must not cost more than £...

Working environment – for example a theatre costume has to be suitable for quick changes

Quality tests and checks that will ensure that criteria measure up to performance requirements

MANUFACTURING SPECIFICATION

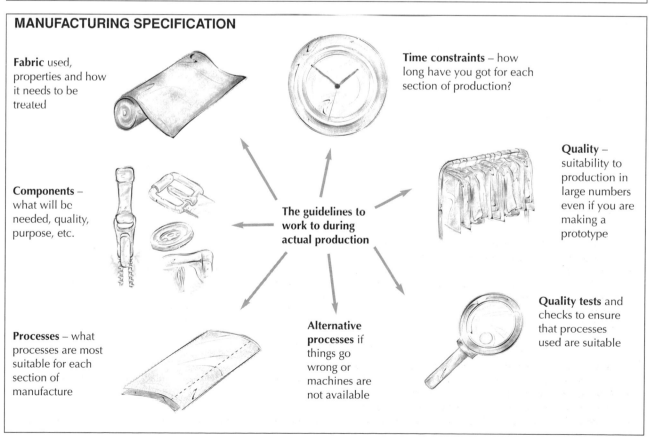

Fabric used, properties and how it needs to be treated

Time constraints – how long have you got for each section of production?

Quality – suitability to production in large numbers even if you are making a prototype

Components – what will be needed, quality, purpose, etc.

The guidelines to work to during actual production

Processes – what processes are most suitable for each section of manufacture

Alternative processes if things go wrong or machines are not available

Quality tests and checks to ensure that processes used are suitable

Initial design ideas

Here is an example of how initial design ideas can develop when working within a brief for your project.

Outline brief
Many young children like comfort blankets to hold when they go to sleep or when they are unhappy or distressed. Unfortunately these can be cumbersome, get dirty quickly and can be unsightly to take on holiday, etc. [**Note:** The overarching topic/theme could have been 'Protection'.]

Remember:
Who? – young children
Where? – bedroom, travelling, etc.
What? – a comforter

Using A3 or A4 paper and cut-outs from magazines, leaflets, etc., quickly make a paper collage for each aspect: faces of young children, children in bedrooms, travelling, etc., and finally children with comforters, soft toys, etc. Pin them up so that you can keep your eye on the main points.

STEP 1

Are commercial fabric comforters on the market for young children? If they are, (1) I should have a good look at them and (2) how can I make mine different? If they are not available, can I produce a possible prototype?

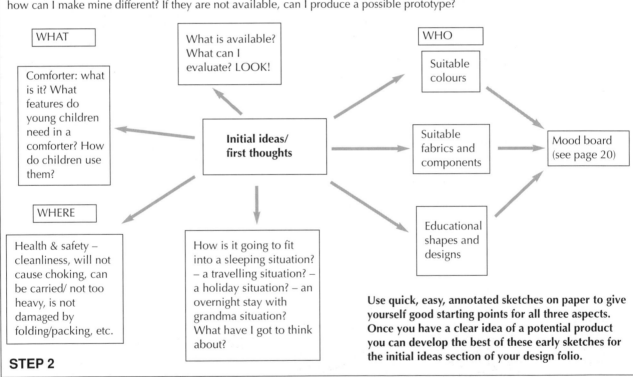

WHAT

Comforter: what is it? What features do young children need in a comforter? How do children use them?

What is available? What can I evaluate? LOOK!

WHO

Suitable colours

Initial ideas/ first thoughts

Suitable fabrics and components

Mood board (see page 20)

WHERE

Health & safety – cleanliness, will not cause choking, can be carried/ not too heavy, is not damaged by folding/packing, etc.

How is it going to fit into a sleeping situation? – a travelling situation? – a holiday situation? – an overnight stay with grandma situation? What have I got to think about?

Educational shapes and designs

Use quick, easy, annotated sketches on paper to give yourself good starting points for all three aspects. Once you have a clear idea of a potential product you can develop the best of these early sketches for the initial ideas section of your design folio.

STEP 2

Your brief has therefore developed to this:
I have found little initial evidence of a variety of fabric comforters available on the market. I am therefore going to design and make a prototype fabric comforter suitable for use by young children when at home or away from home. It must therefore be suitable for carrying by the child or for packing in a suitcase or bag.

Have I missed anything that should now be included?

Now consider:

Which of my quick sketches shall I develop further for my initial ideas?

How can I best represent my initial design ideas clearly and what information must go in the annotation?

STEP 3

Presenting design ideas

Design ideas must always be presented very clearly with good annotation to describe materials, components and proposed processes.
Here is a range of ideas you can use:

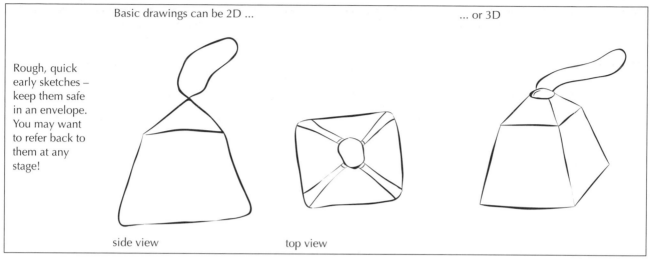

Basic drawings can be 2D ...   ... or 3D

Rough, quick early sketches – keep them safe in an envelope. You may want to refer back to them at any stage!

side view   top view

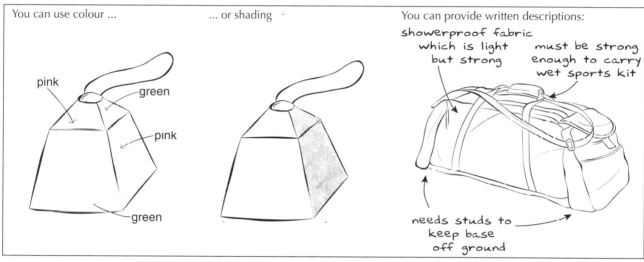

You can use colour ...   ... or shading   You can provide written descriptions:

pink
green
pink
green

showerproof fabric which is light but strong

must be strong enough to carry wet sports kit

needs studs to keep base off ground

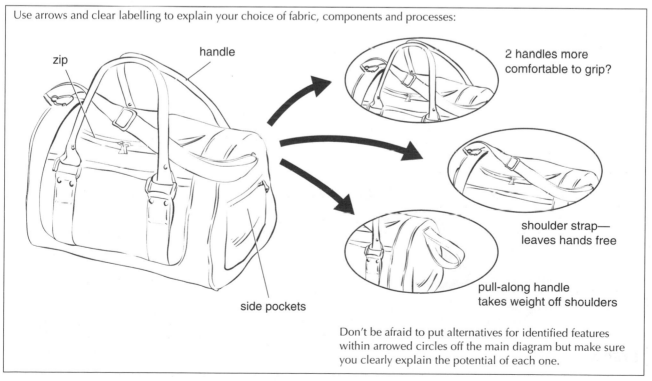

Use arrows and clear labelling to explain your choice of fabric, components and processes:

zip   handle

2 handles more comfortable to grip?

shoulder strap— leaves hands free

pull-along handle takes weight off shoulders

side pockets

Don't be afraid to put alternatives for identified features within arrowed circles off the main diagram but make sure you clearly explain the potential of each one.

Testing fabrics

A number of simple tests can help you identify suitable properties to meet the specification for the product.

A **cloth count** (number of warp and weft threads per unit length) identifies the tightness and looseness of the weave.

Test **wear** by rubbing sandpaper over each fabric piece in the same direction, with equal pressure, for an equal number of times.

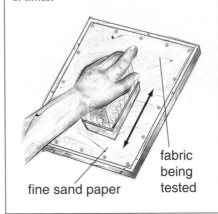

fine sand paper

fabric being tested

Stain resistance: use a pipette to drop liquid stains on fabric samples. Allow to dry, then wash.

Stretch test: use weights on equal-size fabric strips, against a grid. After 1 hour mark fabric position, then remove weights. After a second hour, mark fabric position. This shows fabric stretch and return.

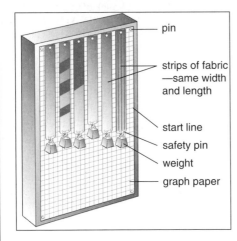

pin

strips of fabric —same width and length

start line
safety pin
weight
graph paper

Insulation: wrap identical beakers with different fabrics. Fill each with boiling water and then plot temperature over time, taking the temperature every 15 min. Always keep a control beaker.

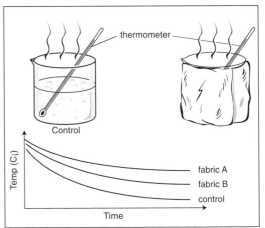

thermometer

Control

Temp (Ci)

fabric A
fabric B
control

Time

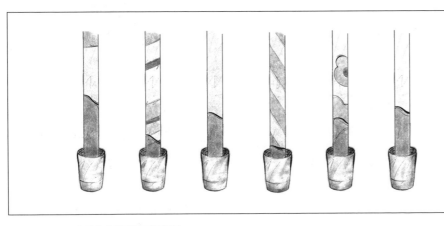

Water absorption: with equal fabric strips and pots, place a pot of water under each strip so that the fabric dips in the water. Record how quickly the water is absorbed up each strip.

Colourfastness and **shrinkage:** use two identically sized fabric pieces for each fabric type. Keep one as a control and wash and dry the other a number of times. Each time check against control for colour and size.

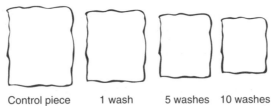

Control piece   1 wash   5 washes   10 washes

Fair testing
For each test, identify what has to stay the same for each fabric type to ensure that your results reflect fair testing procedures.

Evaluating design ideas objectively

When evaluating design ideas you must be **objective**. This means you must base your evaluation on actual results. Your evaluation cannot be biased towards your own opinion or feelings. For example, if you tested a prototype for decorated cushions and you preferred the ones with the appliqué but the majority of your testers preferred the machine-embroidered cushions, you mustn't choose the appliqué ones just because *you* like them!

Unobjective Una

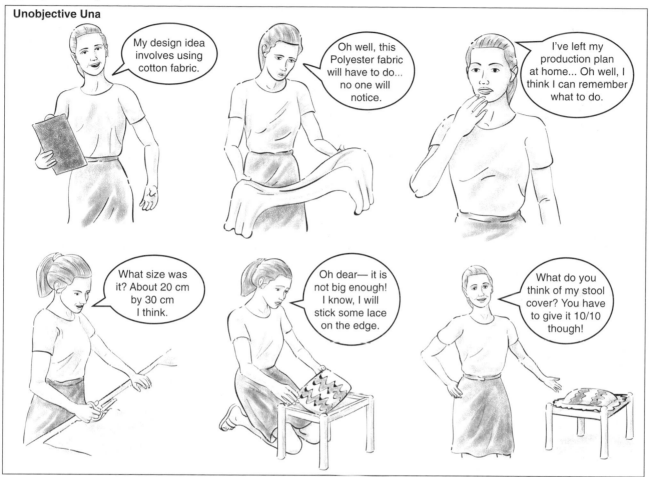

Objective Olivia does things differently.

- Checks and measures accurately
- Follows production plan carefully
- Uses control checks and Gantt chart
- Notes any problems or difficulties that arise

▲ Has planned how she will evaluate her design ideas
▲ Has asked her testers beforehand
▲ Has prepared evaluation forms for her testers to complete

■ Looks at all the results carefully
■ Totals marks and collates information
■ Prepares an objective evaluation of her design idea

REMEMBER

After testing and evaluating a prototype you *might* decide the idea should be rejected. If this is the case, explain *why* it has been rejected.

Modifying design ideas

To modify means to make slight changes or adaptations to your original design idea as it develops. You should start by producing an early prototype product out of cheap fabric (see page 31). This early stage may throw up a number of small problems that you had not considered.

At every stage you should constantly be referring back to your original specification and testing your prototype product against it. If specifications are not going to be met, what can you do?

The answer is simple – consider alternative ideas.

Fabric paint bleeds under the stencil making the edge smudgy.
• Can you add an additional decorative feature around the edge of the design, e.g. beadwork?

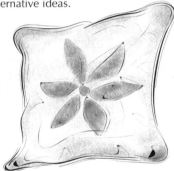

Fabric costs for your large product will be expensive.
• Cheaper alternative fabric but compromise on ideal properties.
• Smaller product – e.g. make a small cushion for the bed rather than a floor cushion.

You shortened the hemline and now it is too short!
• Can you add a contrasting band of colour to the bottom?
• Can you add a frill?
• Can you use the same fabric in a pleated design around the base?

The only overlocker has broken down.
• Can you use an alternative seam method?
• Can you alter your production plan so that you can do something else today and use the overlocker next lesson?

Time is not on your side!
• Can you use machine embroidery instead of hand embroidery?
• Can you alter your initial idea to make it simpler?

Using IT skills

It is very important throughout your project that you show clear evidence of your use of IT. In textiles, IT can be used in a number of ways.

Using graphics software

Scan in a silhouette or basic product shape and save it as an image.

Drop the image into a graphics package and use it as a basis for development.

Use the graphics package to develop ideas for colour and pattern, then flood fill areas of your design.

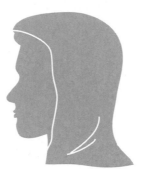

Use the graphics package to develop a background for your design. It could be based on a scanned-in photograph.

Finally drop your completed design onto the background and experiment with fonts and layouts to get the effect you want.

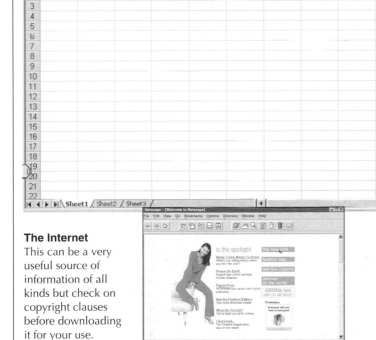

Spreadsheets and databases

These are excellent ways of collating and analysing information.

The Internet

This can be a very useful source of information of all kinds but check on copyright clauses before downloading it for your use.

Producing a production plan

Definitions
A **schedule** is a list of products which a manufacturer plans to produce in a given time.
A **production plan** identifies the main stages of making and the critical control points.

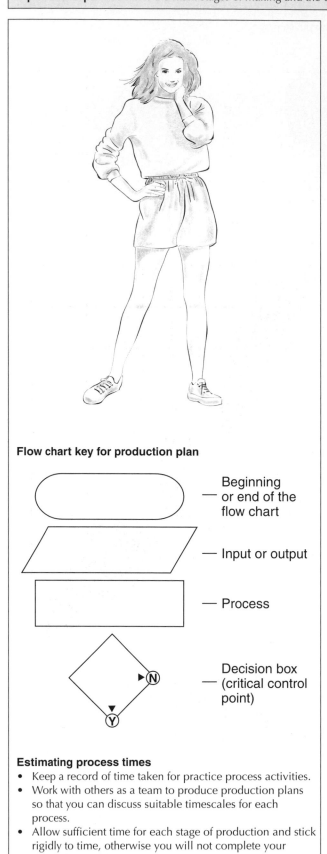

Flow chart key for production plan

⬭ — Beginning or end of the flow chart

▱ — Input or output

▭ — Process

◇ ▶Ⓝ ▼Ⓨ — Decision box (critical control point)

Estimating process times
- Keep a record of time taken for practice process activities.
- Work with others as a team to produce production plans so that you can discuss suitable timescales for each process.
- Allow sufficient time for each stage of production and stick rigidly to time, otherwise you will not complete your product.
- Build in critical control checks and time for alternatives if you meet problems.

A production flow chart for the assembly of pairs of shorts

(Quantity of shorts required)

▱ • fabric pieces • thread • drawstring • energy

3 mins	Join front and back seams
3 mins	Join side seams
1 mins	◇ Check seams Ⓝ
	Ⓨ
7 mins	Machine hem on each leg
7 mins	Machine elastic casing
1 mins	◇ Check machining Ⓝ
	Ⓨ
3 mins	Insert drawstring
3 mins	Press
2 mins	◇ Final check Ⓝ
	Ⓨ

▱ Completed shorts

(Time taken 30 mins)

Producing a Gantt chart

Definition
A **Gantt chart** identifies the main stages in production and charts progress of work.

Across the top, heading the columns, are the days or calendar dates.

The major activities are listed down the left-hand side of the chart.

The thick vertical line indicates the present status of the work.

Use string and a piece of Blu-Tack to make a moveable line!

A Gantt chart for production cushions

■ Planned work ▨ Completed work Activity	Week 1	Week 2	Week 3	Week 4	Week 5	Week 6
Cut fabric pieces	▨					
Cut wadding	▨					
Print cushion fronts	▨▨					
Quilt cushion fronts		▨				
Add appliqué pieces			▨			
Add embroidered features			■			
Insert zip in back pieces				■		
Make frill for edge				■		
Apply frill to front					■	
Apply back to front					■	
Finish, press and inspect						■

Week 4 – starting to fall behind because of machine problems

Make sure that your chart is pinned up and easy for you to see so that you can keep track of where you are and where you should be.

Produce a smaller copy for your project folio and explain how well you were able to keep to it.

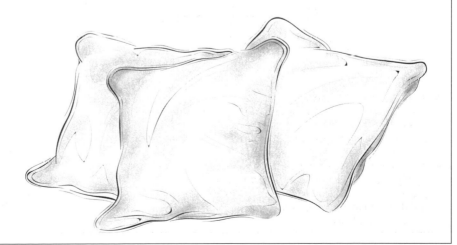

Risk assessment – identifying critical control points

There are **three** main areas to consider:

INPUT

Materials and components

Equipment/processes

All materials and components you plan to use should be thoroughly checked before use.

Faulty equipment can tear fabric and reduce quality. Similarly, it is important to know which processes will work best.

Fabric tests and eye checks over fabric, test components, etc.

Test all equipment and processes before use.

WORK IN PROCESS

Each stage in production

Identify each main stage in the production process:

What has the potential for going wrong?

What is acceptable for a quality outcome?

How can you test/compare for the criteria you have set?

Have you produced swatches/samples for comparison?

Eye checks

Checks for fit

OUTCOME

Finished product

What checks should you carry out here for finish?

Threads

Processes

Fit

Drape

Pressing

Decorative details

Eye checks

Checks for fit

Checks against samples

Trial the product in use

REMEMBER

Build in feedback systems at each point so that you go back to the correct stage if you have a major problem during production. Consider alternative methods so that you know what else you can try if the first method fails to produce the expected results.

For example, if stencil work on fabric results in the paint bleeding around the edge of the stencilled shape, do you waste that fabric or is there a process or technique that could overcome the problem?

Also use a toile or model in cheap fabric to trial and test the design fully before production begins!

Making successful textile products – processes and techniques

Throughout your course for GCSE Textiles Technology you will have been building up your experience in all the skills and knowledge of fabrics, processes and techniques. When designing any textile product, including garments, there are certain areas you need to consider. You must show evidence of how competent you have become when using fabric as a design material.

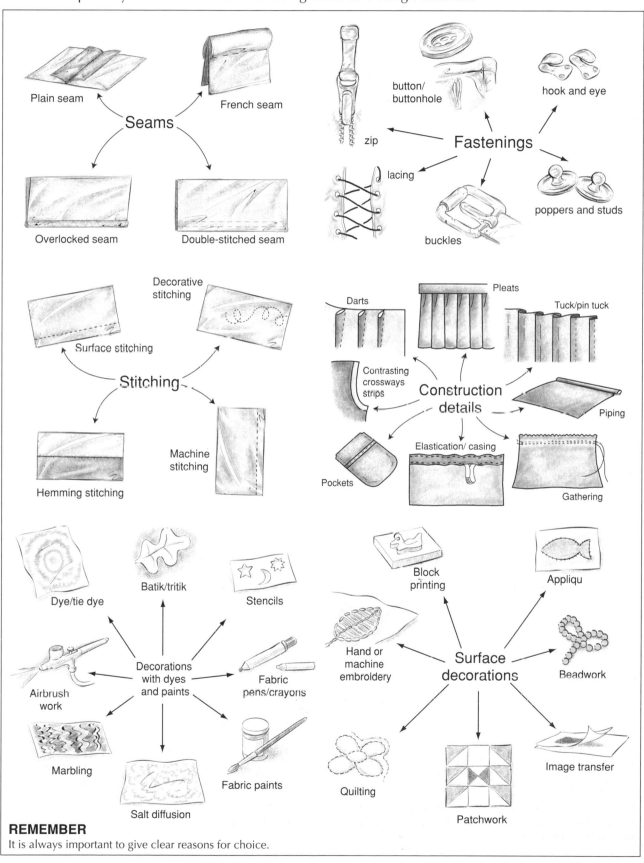

REMEMBER

It is always important to give clear reasons for choice.

Making successful textile products – equipment

While you will be limited to using equipment that is available to you in the classroom or at home, you must show evidence of a knowledge of large-scale production.

USING EQUIPMENT DURING THE PRODUCTION OF YOUR TEXTILE PRODUCT

There are two key points to remember:

- Always use the correct tool/piece of equipment for the job.
- Always practise with the equipment before using it on fabric or the product you are assembling. You will then clearly know how to use it and the limitations of its use.

Do I ...

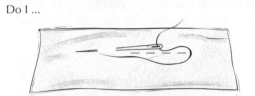

... hand sew?

When joining the seams, do I:

machine the seam and zigzag sew the seam edges together?

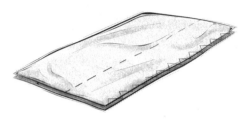

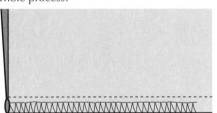

... or machine?

... or do I use an overlocking machine for the whole process?

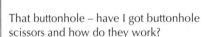

I want to create a design on my fabric. Do I:

create a stencil and use fabric paint to sponge on the design?

That buttonhole – have I got buttonhole scissors and how do they work?

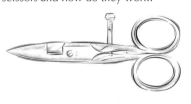

... or have I got access to marbling equipment?

Should I use sharp embroidery scissors or snippers?

Having tested the equipment to make sure that you have chosen the correct equipment for the process, explain in your project **what** you have chosen and **why** it is suitable.

Is it safe?

An important aspect of Textiles Technology is that you consider the safety of the product when in use, and your own safety in making it.

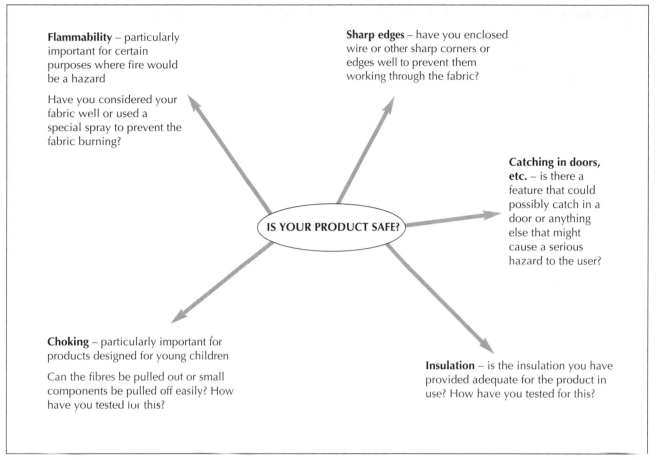

Flammability – particularly important for certain purposes where fire would be a hazard

Have you considered your fabric well or used a special spray to prevent the fabric burning?

Sharp edges – have you enclosed wire or other sharp corners or edges well to prevent them working through the fabric?

IS YOUR PRODUCT SAFE?

Catching in doors, etc. – is there a feature that could possibly catch in a door or anything else that might cause a serious hazard to the user?

Choking – particularly important for products designed for young children

Can the fibres be pulled out or small components be pulled off easily? How have you tested for this?

Insulation – is the insulation you have provided adequate for the product in use? How have you tested for this?

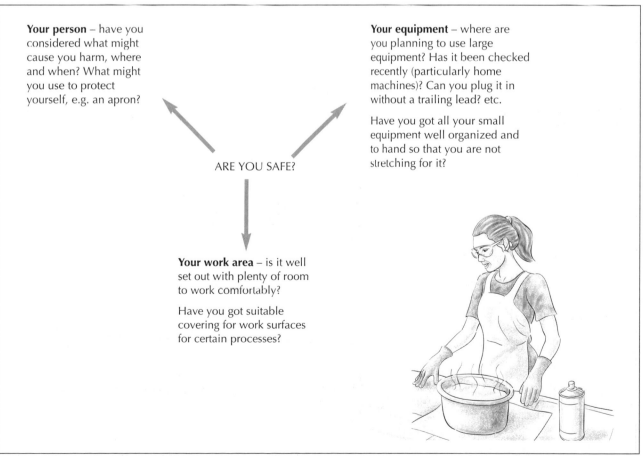

Your person – have you considered what might cause you harm, where and when? What might you use to protect yourself, e.g. an apron?

Your equipment – where are you planning to use large equipment? Has it been checked recently (particularly home machines)? Can you plug it in without a trailing lead? etc.

Have you got all your small equipment well organized and to hand so that you are not stretching for it?

ARE YOU SAFE?

Your work area – is it well set out with plenty of room to work comfortably?

Have you got suitable covering for work surfaces for certain processes?

Making a quality product

There are two main aspects of the quality of a textile item – the quality of design and the quality of manufacture.

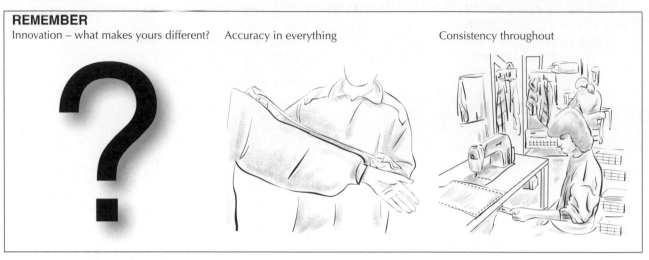

FINISHING TOUCHES TO ENSURE QUALITY

Inspect for loose ends of thread, snags, unfinished seams, etc.   Press the finished item carefully and well!

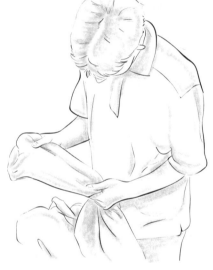

If it is a garment – hang it up correctly or fold it neatly.   Cover it well to protect the product from soiling and damage.

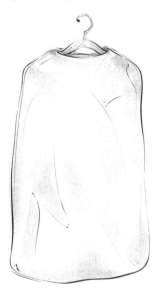

Revision questions (pre-examination)

These questions are designed to help you test your knowledge and understanding of Textiles Technology. They do not represent examination questions. They will make good practice prior to trying examination questions.

1. Name two animal fibres and two vegetable fibres.
2. Explain the use of tailor tacking.
3. Describe two simple methods of hemming fabric.
4. What is an overlocker used for?
5. What is a hot notcher?
6. What is meant by 'intellectual property'?
7. Suggest ways in which packaging can help display the product.
8. Name four different types of collars.
9. What is a microfibre?.
10. Explain micro-encapsulation.
11. What is the linear density of a fibre?
12. Describe the main differences between wet, dry, and melt spinning.
13. Suggest three functions of packaging.
14. What is the difference between a staple fibre and a continuous fibre?
15. Explain the importance of the spinneret to the fibre produced through it.
16. Define 'addition polymerization'.
17. Identify suitable items to help you create a mood board.
18. Identify three different methods of producing bonded fabrics.
19. Explaining the action of mothproofing fabric.
20. How can you test for wear in a fabric?
21. What quality control checks are carried out on fabric before product manufacture?
22. Explain the main difference between warp and weft knitting processes.
23. Define 'spatial relationships'.
24. Name two mechanical finishes and explain what they do to the fabric.
25. Explain the term 'street fashion'.
26. Identify two other possible influences on fashion designers.
27. How do you use a storyboard?
28. There are two main parts to a product specification: identify both of them and explain the main differences between them.
29. What is 'bespoke production'? Explain two different types of bespoke businesses.
30. The seam type most often used is a plain seam. Identify and describe three methods of neatening the seam allowances.
31. Name six different textile components.

32. What is a Gantt chart?
33. Define 'critical control points'.
34. Define 'batch production'.
35. Identify the main ways in which you can prevent hazards in the workplace.
36. Name one fibre which absorbs moisture and explain why it does so.
37. What is the importance of the number of crimps per centimetre to wool?
38. Describe what happens to silk cocoons after they are harvested.
39. Give three reasons for blending fibres.
40. What are the two main twist properties of yarn?
41. Name two methods of printing fabric.
42. Suggest four benefits of evaluating existing products.
43. Suggest two ways ways in which fabric can be used to enhance or emphasise features.

Sample examination questions

1. Here are four small items of equipment.

(i)   (ii)

(iii)   (iv)

   Name each item of equipment. [4]

2. (a) What is a prototype product? 2]
   (b) You have been asked to produce a prototype textile product for use in kitchens. Give **three** points to include in the design specification for this product. [3]
   (c) What are **three** benefits for the manufacturer in producing prototype products? [3]

3. (a) Name the fibres that are produced from each of these natural sources.

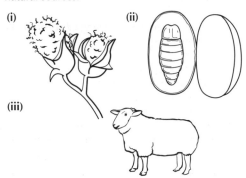

(i)   (ii)

(iii)

                                     [3]
   (b) Name the natural fabric popularly used for heavy winter clothing. [1]
   (c) Give **two** reasons why it is so popular. [2]
   (d) Synthetic fibres are often 'crimped'. Explain **one** benefit this gives to the fabric in use. [2]

4. (a) State what these care symbols mean.

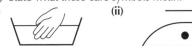

(i)   (ii)

(iii)   [3]

   (b) When washing textile products there are four important points to consider. What are they? [4]
   (c) Give **one** reason why a textile product can 'shrink' when washed. [1]

5. (a) Name the four pattern markings shown.

(i)   (ii)

(iii) — ✂ — (iv) ═══════ [4]

   (b) Explain **two** different methods of transferring these pattern markings onto fabric. [4]

6. A manufacturer wants a logo for a range of textile products for use on the beach.
   (a) Suggest **one** method or technique to work a logo during mass production. [1]
   (b) Explain why this method is suitable. [2]
   (c) Draw **two** design ideas for a logo for beach products. [2]
   (d) Give **one** reason why the logo is important to a manufacturer. [2]

7.

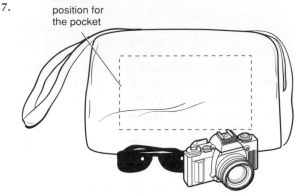

   position for the pocket

   This prototype beach bag needs a large, secure outside pocket to hold small personal items.
   (a) Draw a suitable design for the pocket showing **three** features to protect and secure these items. [8]
   (b) Explain your choice of fastening for the pocket. [2]
   (c) Describe how the pocket would be securely attached to the main part of the bag. [2]

8. The product range has a matching T shirt to protect children from the sun when playing on the beach.
   (a) Give **two** advantages of screen printing a design on the front of the T shirts. [2]
   (b) How is T shirt cotton fabric constructed? [2]
   (c) How does this fabric construction help protect skin from harm by the Sun? [2]

9. (a) What is an overlocked seam? [1]
   (b) Give **three** benefits of using overlocked seams on textile products. [3]
   (c) Name **two** other types of seam. [2]
   (d) Give **one** example of where each type of seam would be used. [2]

10. (a) List **five** points of information which must be displayed on the packaging of textile products. [5]
   (b) Give **two** reasons why this information is important. [2]
   (c) Explain **three** uses of packaging for textile products. [3]

11. (a) Explain where the following fabric properties would be important:
     (i) resistance to sunlight [2]
     (ii) water-repellence [2]
     (iii) resistance to burning [2]
   (b) When a fabric has 'lustre', what does this mean? [2]
   (c) Fleece fabric has become popular.
     (i) Give **one** reason why. [1]
     (ii) Give an example of a product where it is used. [1]

12. (a) State **two** methods you would use to evaluate a prototype product. [2]
   (b) Describe **one** suitable method to test fabric for wear. [3]
   (c) State **four** benefits of evaluating available textile products in use. [4]
   (d) Explain how to test fabric for colour fastness. [2]

Sample examination questions

1. Sewing machines come with many useful features. Explain the advantages of buying a CADCAM sewing machine. [4]

2. (a) What is a prototype product? [2]
   You have been asked to produce a prototype range of textile products for kitchen use.
   (b) List a design specification for such a product range. [4]
   (c) What has to be considered when deciding on a suitable price range for the products? [2]

3. (a) Explain the meaning of 'natural fibres'. [2]
   (b) A winter product range of textile items requires a fabric which provides warmth. Compare and contrast a suitable natural fibre with a suitable synthetic fibre. [4]
   (b) Synthetic fibres can be 'crimped'. What are the benefits of this process for the fabric produced? [2]

4. (a) Explain what these care symbols mean.
   (i)   (ii)   (iii)   [3]
   (b) Describe the **four** points to be considered when laundering a textile product. [4]
   (c) Explain what happens when a textile product 'shrinks' when laundered. [1]

5. (a) Identify **four** pattern markings and in each case explain their use. [4]
   (b) With the aid of diagrams, describe **two** main methods of transferring pattern marking onto fabric. [4]

6. A manufacturer wants a logo for a range of textile products for use on the beach.
   (a) Explain **one** suitable method for mass production of a logo design. [1]
   (b) Draw **two** design ideas for the logo and in each case identify the correct sequence for production. [4]
   (c) Explain the importance of the logo to the manufacturer and how it can be protected. [2]

7.
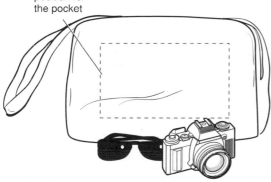
   position for the pocket

   This prototype beach bag needs a large, secure outside pocket to hold small personal items which must also be detachable from the main body of the bag.
   (a) Draw a suitable design for a detachable pocket showing a number of features to protect and secure small personal items. [8]
   (b) Explain your choice of fastenings both inside and outside the product. [2]
   (c) When attached to the main body of the bag, explain how you would design in features to ensure it remained secure in use. [2]

8. The product range has a matching T shirt to protect children from the Sun when playing on the beach.
   (a) Describe fully the process of screen printing a design on the front of the T shirt and the advantages of using this process. [4]
   (b) Explain why the fabric aids protection from the harmful rays of the Sun. [2]

9. (a) What is an overlocked seam? [1]
   (b) Overlocked seams are frequently used on manufactured products. Explain the benefits of the use of this type of seam. [3]
   (c) Describe two other methods of seam production, giving examples of their use in products with reasons for your choice. [4]

10. Packaging plays an important role in the sale of textile products.
    (a) List the information found on textile product packaging, in each case explaining the importance of the information for the manufacturer and/or customer. [5]
    (b) What are the main uses of packaging for the product? Explain, giving examples of the types of material used for each purpose. [3]
    (c) What are combi clips? [2]

11. (a) Explain where the following fabric properties would be important.
    (i) resistance to sunlight [2]
    (ii) water-repellence [2]
    (iii) resistance to burning [2]
    (b) Some fabrics have 'lustre'. How is this effect created on a fabric? [2]
    (c) Explain why fleece fabric has become so popular for use in both garments and textile products. [2]

12. You have completed a prototype product.
    (a) Explain how you would evaluate this prototype product fully and the importance of this for manufacturing. [3]
    (b) Describe how the fabric used could be tested fully for
    (i) wear and tear
    (ii) colour fastness [4]
    (c) Identify four critical control points on an assembly line for a textile product. [4]

Revision answers

1. Animal fibres: silk, wool; vegetable fibres: cotton, linen.

2. To transfer pattern markings onto fabric pieces for assembly.

3. **Slip hemming**: turn up the required hem depth and tack along the fold at the base. Turn under the top edge of the hem and edge stitch if required. Slip stitch the hem edge to the fabric and remove the tacking. **Narrow machine hem**: turn up a 8 mm hem depth and press. Turn under the top 4 mm of the hem and pin into position. Machine along the edge of the hem close to the fold.

4. An overlocker machine machines the seam, trims it, and neatens it as one process.

5. A hot notcher is used in industry to mark fabric pieces for assembly.

6. Intellectual property is the right of a designer to protect designs developed by them under a trademark brand name.

7. Packaging can have a built-in method of displaying the product on a stand, can display specific features of the product to the customer and can hold the product in a specific shape for display.

8. Shawl, shirt, revere, mandarin, roll, etc.

9. A microfibre is an extremely fine synthetic fibre less than 1 denier thick.

10. Micro-encapsulation is a hollow fibre into which crystals can be placed which break down over time allowing perfumed clothes, vitamins to be absorbed by the skin, house dust mites to be killed, etc.

11. Linear density is the weight of a fibre in proportion to its length.

12. The differences lie in:
    • whether the polymer used is already a solution or has to be melted
    • the size and the shape of the spinneret holes
    • the way the fibre is hardened, in a bath of solution or in air

13. Packaging can protect from soiling and damage, can display a product and can help maintain shape in a product.

14. A staple fibre is a short fibre, a continuous fibre is a long fibre.

15. The spinneret determines the thickness and shape of the fibre produced.

16. Addition polymerization produces long polymers of fibres made from the same basic unit.

17. For a mood board you need fabric pieces, wools and threads to denote suitable colours and patterns, photographs to aid scene setting, braids, ribbons and other components to help identify additional features, etc.

18. Bonded fabrics can be produced by the addition of adhesives, a solvent to stick fibres together, heat to melt fibres together or by stitching fibres together.

19. Chemicals are applied to the fabric which help repel moths.

20. To test for wear you need to establish a fair test in which fabrics are subjected to the same number of rubs with an abrasive surface, in the same direction and with the same force so that you can compare the results.

21. Fabric is weighed to ensure that it is correct, then eye check for tears, pulls, holes, colour/pattern defects, etc.

22. Warp knitting: the loops are created vertically up the knitting; Weft knitting: the loops are created in horizontal rows. Weft knitting unravels easily and will ladder or run if cut. It is stretchy and has a right and a wrong side. Warp knitting is faster and cheaper to produce, only works with filament yarn and keeps its shape well.

23. Spatial relationships means the way we see products in relation to the things around them.

24. Raising is a process which teases out fibres to create a pile on a fabric. Calendering involves sets of rollers and a combination of heat and pressure to improve lustre and emboss fabrics.

25. Street fashion is fashion design that evolves within youth culture before being taken up and developed by designers.

26. Two from natural sources, man-made objects, history and theatre/film culture, etc.

27. A storyboard helps you plan out the main stages of manufacture/assembly for a textile product.

28. Design and manufacture. A design specification considers what needs to be built into the design of the product to make it suitable for the purpose. A manufacturing specification identifies the quality required during manufacture and assembly to ensure that the product is well made.

29. A bespoke business is one which produces one-off items or very small numbers of products of the same type. A tailor produces made-to-measure suits for men and women; a craft outlet produces small quantities of products to meet a specific area of the tourist market.

30. Methods include turned and stitched edges; bias bound edges; pinked and stitched edges and zigzag stitched.

31. Stud fastenings; lace; buckles; zips; iron or sew-on badges; chains.

32. A Gantt chart identifies the main stages of development for a product and charts progress of work.

33. Critical control points are places in production where it is essential to carry out checks before allowing production to continue.

34. Batch production is the production of products to specific quantities in a production run.

35. Hazards can be prevented by maintaining a tidy, well organized work area, storing tools correctly, using protection for body, eyes and surfaces as needed and using equipment correctly with a guard if appropriate.

36. Wool absorbs moisture because the protein bundles are able to move apart, breaking the bonds between them. These bonds reform as moisture is lost.

37. Crimps in a fibre trap air for insulation. The greater the number of crimps per centimetre, the better the insulation properties.

38. The silk cocoons are softened in hot water so that the silk ends can be found. The silk threads from a number of cocoons are then twisted together to form a fibre for weaving.

39. Fibres are blended to reduce costs of more expensive fibres, to combine the best properties of the fibres and also to add lustre to duller fibres.

40. The direction of the twist and the number of twists per unit length. A high twist produces a smooth, dense and expensive yarn, while a low-twist yarn is rough with a greater volume and a low cost.

41. Fabric can be printed using screen printing and flat printing methods and also the new inkjet printing method.

42. How others have solved design problems, compare effectiveness of different techniques, compare different fabrics, compare prices and likely costs.

43. Horizontal stripes emphasise width, and vertical stripes emphasise height.

Sample examination answers

FULL/SHORT COURSES – FOUNDATION

1. 1 mark each for dressmaker's shears, tracing wheel, seam ripper, tailor's chalk.

2. (a) 2 marks for a full account to include: detailed, accurate working model/copy/trial/mock-up/sample of design. Either full size or scaled/made in cheaper fabric.
   (b) • Fashionable/attractive/colourful
   • Maximum and minimum dimensions
   • Name of fabric
   • Insulation where appropriate
   • Finished – stain/water resistant
   • Details of fastenings/pockets/components
   • Non-flammable
   • Cost
   1 mark each for any 3 of the above or any other appropriate specification.
   (c) • Test performance of working model
   • Test for suitability to purpose
   • Seams, fastenings, strength, stitching size and shape
   • Use for market research
   • To avoid problems in production.
   1 mark each for any 3 of the above or any other appropriate benefit.

3. (a) Cotton, silk, wool – 1 mark for each.
   (b) Wool
   (c) • Fibre structure traps air which is a good insulator
   • Close weave of fabric again traps air
   • Wool repels surface water because of natural grease
   1 mark each for any 2 of the above reasons.
   (d) 2 marks for a detailed explanation to include: the crimping process allows fibres to trap air which is a good insulator; the more crimps per cm of fibre, the better the insulation properties of the fabric.

4. (a) (i) Hand wash only.
   (ii) Cool iron.
   (iii) Do not tumble dry.
   1 mark each.
   (b) 1 mark each for temperature, mechanical action, washing medium (i.e. detergent dissolved in water) and time.
   (c) Textile products shrink when the fibres become entangled in each other due to mechanical action, or they shorten because of high temperatures.

5. (a) (i) To the fold.
   (ii) Notches/balance marks.
   (iii) Cutting line.
   (iv) Lengthen or shorten the pattern.
   1 mark each.
   (b) 2 marks for each method:
   – tailor tacking: stitch through the pattern markings and fabric leaving loops, snip between the loops, gently pull the layers of fabric apart and cut the threads between them.
   – carbon paper and tracing wheel: place the carbon paper between the fabric and the pattern, use the tracing wheel to trace along the pattern marking so that the carbon marks the fabric.

6. (a) 1 mark for CADCAM embroidery using a POEM machine or a computer-controlled embroidery machine.
   (b) 2 marks for an explanation that includes: computer control means accurate sequence of processes to produce a consistent result each time. It is also fast and cost effective.
   (c) 1 mark for each suitable design idea: shells/sandcastles/sea/sun/boats/surfers, etc.

(d) A logo is important because it identifies products with the manufacturer and can be a trade mark.

7. (a) 2 marks for a well drawn idea plus 2 marks each for 3 features such as:
   • internal pockets/sections of different sizes
   • padding/quilting for protection
   • inner fastenings to hold items in place
   • expanding gussets at the side
   The final drawing should identify these features as well as how the pocket will be attached to the bag, external fastening, etc.
   (b) 2 marks for an explanation that includes: use of a quick and easy method that is secure in use but will allow for some expansion for bulkier items, e.g. a modern buckle system.
   (c) 2 marks for a description that identifies the need for reinforcement sewing at the corners to prevent the weight of the items pulling the pocket away from the main part of the bag.

8. (a) 1 mark each for:
   • the design can be built up with a small number of colours
   • the design can be repeated easily on any number of T shirts
   (b) Knitted.
   Brief description of warp knitting.
   (c) 1 mark each for:
   • closeness of the knitted loops prevents the rays of the Sun from passing through
   • thickness of the fibre also helps

9. (a) An overlocked seam is produced by an overlocking machine which sews the seam, trims and neatens the seam allowance at the same time.
   (b) 1 mark each for:
   • quick
   • one operation
   • suitable for curved seams
   (c) 1 mark for 2 of:
   • open or plain seam
   • double-stitched seam
   • French seam
   (d) 1 mark for each suitable application:
   • open seam – summer cotton clothes, cushions
   • double-stitched seam – trousers, jackets, curtains
   • French seam – children's clothing, products that need to stretch over items

10. (a) 1 mark each for 5 of:
   • fabric content
   • size details
   • description of product
   • cost
   • colour
   • manufacturer information
   • country of origin
   • barcode
   (b) 1 mark each for each appropriate reason, e.g.
   • size is important for garments and products to fit specific situations e.g. curtains
   • fabric content identifies properties of the fabric in use
   (c) 1 mark each for 3 of:
   • protect from soiling
   • protect from damage
   • to maintain shape
   • to advertise the product
   • to display the product

11. **(a)** **(i)** Where a fabric was in bright sunlight for long periods of time – e.g. curtains, summer garments.
(ii) Where water could result in wetness or damage, e.g. macs and coats, bathroom or kitchen items.
(iii) Where burning fabric could result in danger or damage to other things, e.g. furnishing upholstery, children's nightwear.
(b) 2 marks for an explanation which includes reference to the way the fibres reflect light, the fabric is smooth and shines in light.
(c) **(i)** 1 mark for one of:
- warmth
- lightness
- comfort in wear

(ii) 1 mark for one of:
- jackets
- nightclothes
- headwear/hats
- scarves
- foot warmers
- blankets/covers

or any other appropriate item.

12. **(a)** 1 mark for each method:
- market research
- test in use

(b) 3 marks for an answer which includes: reference to fair testing, a suitable method of rubbing the surface of the fabric, number of rubs, weight of the rub, evaluating the result.
(c) 1 mark for each benefit – four of:
- test for suitability of purpose
- test for customer reaction
- cost/test for suitability of price
- test for wear and laundering etc.
- test in different situations

(d) 2 marks for a full explanation to include:
- control/test piece use
- washing/drying the test piece a number of times
- standing the test piece in sunlight
- reaction of test piece to acid/alkaline conditions
- comparison with control

Sample examination answers

FULL/SHORT COURSES – ANSWERS

1. Advantages:
- develop your own designs for production
- high consistency of quality production
- changes can be made easily
- a very quick process and therefore cost effective
- a full range of stitches/effects can be produced

4 marks for a full answer to include at least 4 of these points.

2. **(a)** 2 marks for a full account to include:
– detailed, accurate working model/copy/trial/mock-up/sample of design. Either real size or scaled/made in cheaper fabric.
(b) Design specification for kitchen textile range:
– fashionable/attractive/colourful
– maximum/minimum dimensions
– fabric identified or properties specified
– insulation provided where appropriate
– finishes required – stain resistant/water repellent where appropriate
– details of fastenings/pockets/components
– non flammable
– cost

4 marks for a well produced list that details 8 points.
(c) Price range will depend on:
– target customer group
– fabrics/components used
– type of kitchen/home range targeted
– potential retail outlet

2 marks for identifying at least 3 points.

3. **(a)** Natural fibres are sourced from nature – 1 point. Vegetable fibres from plants, animal fibres from animals – 1 point.
(b) 2 marks for comparison – at least 2 points.
2 marks for contrast – at least 2 points.
Wool and acrylic
– methods of providing warmth/insulation
– drape/feel/comfort in wear
– laundering/cleaning/care
– costs

(c) Crimped fibres trap air. Number of crimps per cm – the higher the number the better the insulation properties. 2 marks if both these points included.

4. **(a)** **(i)** Hand wash only.
(ii) Cool iron.
(iii) Do not tumble dry.
(b) The four points:
- Temperature must be correct for the fabric.
- Mechanical action can help remove dirt from fibres such as cotton but can result in shrinkage in fibres such as wool.
- Washing medium must be correct for the fibre. Acids/alkalis can damage some fibres. Biological detergents can break down natural fibres such as wool.
- Time – over a long time/too much soaking can result in some fibres shrinking but strong fibres such as cotton can benefit from more prolonged washing.

1 mark for each explanation.
(c) The fibres either become entangled together or the fibres melt in high temperatures.

5. **(a)** To the fold; straight grain, dart; lengthen/shorten; cutting line; seam line; notches/balance marks.
1 mark for each of up to 4 correct ones.
(b) 2 marks for well labelled diagrams that show the process (1 mark each method).

2 marks for description (1 mark each method).
Methods – tailor tacking
– carbon paper/tracing wheel
– tailor's chalk/fabric pen

6. **(a)** 1 mark for brief explanation of CADCAM embroidery.
(b) 2 marks for design ideas (1 mark each).
2 marks for sequence of stitches/colours (1 mark each).
(c) • identification of products with the manufacturer
- trade mark/intellectual property rights to the designs

1 mark for each point.

7. **(a)** 2 marks for a suitable pocket drawn well and labelled
4 marks for internal pocket features shown such as sections or internal pockets/with fastenings/with padding or quilting where appropriate.

2 marks for a suitable method of attaching the pocket to the main body of the bag so that it can be removed and used separately but remain secure in use.
– zip
– buckles
– studs

(b) Variation in fastenings to suit purpose.
e.g. Velcro to hold sections closed; zips to hold pockets (internal) firm; studs to detach/attach items etc.
2 marks where understanding of fastenings and how they can best be used are shown.

(c) Secure when attached – looking for additional features to hold the pocket firmly in place, e.g. zip to attach it to the body of the bag with strong studs at the lower corners/buckles and straps system to ensure the pocket is secure at all points.
2 marks for a suitable method well described.

8. **(a)** 4 marks for a description that includes:
- development of screen design
- application of fabric colour/paint
- build up of colours – lightest to darkest/drying between colours
- advantages of use for mass production to include reuse of screens/cost etc.

(b) The size/nature and shape of the fibre helps deflect the rays.
The closeness of the warp knitting process helps prevent rays passing through.
1 mark for each point.

9. **(a)** An overlocked seam is produced by an overlocking machine which sews the seam, trims and neatens the seam allowance at the same time.

(b) Benefits of overlocking:
- quick and easy all-in-one process
- machine made/consistent manufacturing process
- can be used on curved/straight seams
- can be used on a variety of fabrics
- low-cost process for the manufacturer
3 marks for identifying these 5 points.

(c) 2 marks for each seam type to include 1 mark for describing the method and 1 mark for a good example with reason for choice, e.g. description of a French seam and its suitability for shear fabrics/children's clothes.

10. **(a)**
- Size – essential for garments and for size-specific products such as tablecloths, bedding and curtains.
- Fabric content – important by law to give accurate fibre content for the product as it indicates potential properties for the product in use.
- Price – important for the customer to be able to compare products and prices.
- Description – can include photograph of the product in use to explain to the customer the type of situation the product has been designed for.
- Barcode – manufacturer/retailer – to identify how many of the products are being sold for restocking.
- Manufacturer's name/logo/details – to allow the customer to see who has made the product and where it has been made.
Up to 5 marks for 5 well explained points.

(b) Main uses:
- protection from soiling/damaging – plastic/card/paper
- shape – card/paper/moulded plastic/pins
- display – card/plastic moulded as hooks
- advertise – paper photographs
3 marks for 3 well explained points.

(c) A combi clip releases red ink if tampered with; it is a security device to prevent shoplifting.

11. **(a)** **(i)** Where a fabric is in bright sunlight for long periods of time – curtains, furnishing, summer garment.
(ii) where water could result in wetness or damage – rainwear/coats, bathroom/kitchen items.
(ii) where burning fabric could result in danger or damage – furnishing fabrics or chairs/cushions/bedding; children's nightwear.
2 marks each.

(b) Lustre – pressure and temperature used to create smooth fibres/fabrics which reflect light.
2 marks for an explanation to include these points.

(c) Fleece fabric is warm due to the ability to prevent body heat being lost, but extremely lightweight due to the fineness of the fibres. It has a soft, comfortable feel. It is used in garments for jackets, suits, nightwear but also for covering hot water bottles, tea cosies and anywhere where heat needs to be retained for warmth.
2 marks for a good explanation.

12. **(a)** Evaluation:
- test in use in various situations/durability/dimensions/weight
- test for purpose/range of purpose/distortion in use/washability/handle
- market research with potential customers to look at appeal/cost/potential use
- evaluate against available products to compare/contrast
- processes for manufacture/suitability and ease of assembly
- quality – design/manufacture – identify potential quality control points
3 marks for at least 3 points well explained.

(b) 2 marks for each test to include:
- fair testing
- use of controls
- details of test
- results to be looked for

(c) Critical control points:
- fabric – eye check for tears/pulls etc.
- Cut pieces of fabric – check correct sizes/bundling etc.
- Seam production – check for broken threads, missed stitching/neatening
- Components e.g. fastenings – check open correctly and inserted correctly.
- Final inspection – loose threads/well pressed
1 mark for any 4 identified and explained.

Index